Solange dos Santos Moragas
Vivian Schutz
Luis Carlos Santiago

Bone Marrow Transplant: Health Technology Assessment

Solange dos Santos Moragas
Vivian Schutz
Luis Carlos Santiago

Bone Marrow Transplant: Health Technology Assessment

Cost-minimisation study

Imprint

Any brand names and product names mentioned in this book are subject to trademark, brand or patent protection and are trademarks or registered trademarks of their respective holders. The use of brand names, product names, common names, trade names, product descriptions etc. even without a particular marking in this work is in no way to be construed to mean that such names may be regarded as unrestricted in respect of trademark and brand protection legislation and could thus be used by anyone.

Cover image: www.ingimage.com

This book is a translation from the original published under ISBN 978-613-9-72940-1.

Publisher:
Sciencia Scripts
is a trademark of
Dodo Books Indian Ocean Ltd. and OmniScriptum S.R.L publishing group

120 High Road, East Finchley, London, N2 9ED, United Kingdom
Str. Armeneasca 28/1, office 1, Chisinau MD-2012, Republic of Moldova, Europe
Printed at: see last page
ISBN: 978-620-7-91438-8

I dedicate this achievement to God, for whom are all things, to my parents, in memoriam, to my husband Gesiel and to my children, who have been my motivation to persevere to the end.

ACKNOWLEDGEMENTS

To God, for the opportunity and the trust given at all times,

To my supervisor, Professor Vivian Schutz, for her commitment and professionalism,

To the professors: Antônio Augusto de Freitas Peregrino, Roberto Carlos Lyra da Silva, Luís Carlos Santiago, Daniel Aragão Machado, for their availability and guidance during the construction of this dissertation,

To nurses Rita de Cássia Machado Torres, head of nursing at the National Bone Marrow Transplant Centre, and Ana Cristina Rangel, Supervisor of the Inpatient Unit at the National Bone Marrow Transplant Centre, for their support and encouragement, which contributed greatly to making this dream come true,

To nurse Cecília Borges, for her contribution during the construction of this dissertation,

To nurse Valéria Fernandes pinho, for her collaboration in the initial construction of this project,

To pharmacist Carolina Lopes Martins, for her friendship, contribution and support,

To nurse Luís Antônio Corrêa Motta, for his partnership and generosity during data collection,

To the nursing technician Marcelo Faria Ramos, for his support during difficult times,

Nurse Luísa Viegas Pereira dos Santos, for her invaluable help with data collection,

To my husband Gesiel Lazaro Barbosa, for his constant support and companionship throughout this process,

To my colleagues Edmilson da Silva Ferreira, Adelaide dos Anjos, Francisca Alves da Silva, Elizete Peixoto and Eliane Souza Sampaio, for their understanding and encouragement,

To nurse Evandro G. Fidelis, for his vote of confidence and partnership,

Nurse Marlei for her partnership during difficult times,

To Mr Osmar Peixoto, for his help during data collection,

To the pharmacy and storeroom service of the National Cancer Institute, for their great help in collecting the data,

To my colleague Thiago Melo, for his willingness to help,

To Dr Luís Fernando Bouzas, for his collaboration with the departments, contributing to the collection of data,

To my friend Carlos from Cemo, for his support and encouragement during this journey,

The women of the IADJ, for their support and intercession with God, for the fulfilment of this dream,

To my children Gabriel, Beatriz and Guilherme, inspirational sources in my life.

Cast your bread upon the waters, for after many days you will find it. Ecclesiastes11: 1

SUMMARY

The object of this study is the cost of Allogeneic Stem Cell Transplantation (HSCT) and its general objective is to analyse the financial cost of Allogeneic HSCT in a Federal Hospital in the city of Rio de Janeiro. This is an evaluative and retrospective study, with analysis of medical records, carried out at a Bone Marrow Transplant Centre in Rio de Janeiro. A total of 158 medical records were analysed from 2006 to 2012, finding a sample of 14 patients who had undergone allogeneic, related bone marrow stem cell transplants. The data collected was organised into an electronic database by typing it into Excel 2010 spreadsheets, from where it was exported and presented in graphs, charts and tables. The following cost items were analysed: supplies, medicines, tests carried out on the patient and the time used by nursing staff during care for each year of data collection (2006 and 2012) and for each phase of the transplant (pre-, trans- and post-HSCT phase). The cost of the transplant in the two years studied came from the sum of the cost of material inputs, medicines, exams and the time spent by the nurse in providing care. The decision tree model was used to analyse the cost-minimisation of HSCT, and variables related to infectious complications in transplantation and their respective costs were used to compose it. Results: Comparing the costs between 2006 and 2012, we found higher costs in 2006, when the Systematisation of Nursing Care (SNC) had not been implemented, compared to 2012, when the SNC had already been implemented. The cost-minimisation analysis showed a lower cost for related allogeneic HSCT in 2012 (R$ 431,482.71) compared to 2006 (R$ 574,688.22), suggesting that SNC has helped to reduce the cost of the procedure. The tornado graph showed the difference in costs between the two interventions, with an incremental cost of R$143,205.51 for 2006. Knowing the impact of HSCT costs for the SUS, and the need for more beds available for treatment, its effectiveness must be guaranteed by planning the work of nurses with the SNC.

Keywords: Haematopoietic stem cell transplantation. Cost and cost analysis. Systematisation of nursing care.

SUMMARY

CHAPTER 1

INTRODUCTION

1.1 PROBLEMATISATION

As an oncology nurse who has been working in a bone marrow transplant centre for 20 years, I have been involved in changes related to the nursing care provided in the transplant unit and its possible influence on transplant costs.

In 2004, there was a Hospital Accreditation process, in which the Systematisation of Nursing Care was a requirement. Prior to this implementation, nursing work was based on the biomedical model, with parts of the Nursing Process being carried out (evolution and prescription), but without methodology.

During its implementation, a group of nurses engaged in this process went in search of examples of success in the health market through benchmarking, where the managers of the institution explored were involved so that the systematisation of care could be implemented, with a view to meeting the requirements of Hospital Accreditation, which aims to elevate a Brazilian institution to international levels of excellence that rely on SNC models.

On this subject, the following question can be raised: Can the cost of stem cell transplants be reduced with the implementation of Systematised Nursing Care?

Today, there is a worldwide need to rationalise health costs due to the finite resources of health systems and the growing and ageing population. This rationalisation must have a methodological basis that substantiates the assessment and points to monetary values, as well as the clinical benefits associated with decisions regarding the choice of a particular technology. In this context, the use of Health Technology Assessment (HTA) is recognised as a method that can help managers make decisions about incorporating a new technology. In the field of oncology, for example, the need for better management is a priority due to the high costs of cancer control treatments (NITA, 2010).

According to the International Agency for Research on Cancer (IARC) of the World Health Organisation (WHO), there were 14.1 million new cases of cancer and a total of 8.2 million cancer deaths worldwide in 2012. The forecast for 2030 is 21.4 million new cases and 13.2 million deaths. In Brazil, the estimate for 2015 points to the occurrence of approximately 576,000 new cases of cancer (INCA, 2014).

These epidemiological data point to an increase in demand for care in the oncology area in the near future. This leads us to state that the development of technologies in the field of oncology is one of the most intense in the health production complex (NITA, 2010).

Haematopoietic stem cell transplantation (HSCT), also known as bone marrow transplantation, is a therapeutic procedure in which haematopoietic cells are infused intravenously in order to restore bone marrow destroyed by chemotherapy. This treatment option is accepted when haematological toxicity is the limiting factor or when the pathological process directly affects the bone marrow (NICOLAU, 2004).

As well as being a highly complex procedure, bone marrow transplants also have high costs. In the United States, the average cost, according to Rizzo (1999), was estimated at U$193,000 per case. The main limiting factors are the small number of beds available for the procedure and the high cost of new drugs to treat the patient. There are three types of bone marrow transplant: autologous, syngeneic and allogeneic (NICOLAU, 2004).

In autologous transplantation, the patient receives their own bone marrow, which is, in most cases, a less complex transplant. The syngeneic transplant is between twin siblings and is a rarer modality due to the low frequency of identical twins in the population. In the third modality, allogeneic transplantation, the patient receives bone marrow from another person, who may or may not be a relative. When the donor is a relative, it is called a related allogeneic transplant. When the donor is not a relative, it is called an unrelated transplant. In this modality, the search for a compatible donor takes place through the National Registry of Voluntary Bone Marrow Donors (Redome). Bone marrow transplants are regulated by ministerial decree number 1,217 of 1999. In 2009, the Brazilian National Transplant System joined the National Marrow Donor Programme (NMDP), allowing foreign patients who need a donor to access Brazilian stem cell banks. This measure expands the possibilities of searching for donors through Redome, reducing the costs of international searches for our patients, which, according to the Ministry of Health, amount to 45 per cent of all searches (BRASIL, 2009).

Transplants can also be classified according to the conditioning regimen into: 1-myeloablative, where chemotherapy is carried out in high doses; 2- non-myeloablative conditioning regimen, where the dosage of chemotherapy is reduced, and 3- reduced toxicity regimen, where conditioning uses immunosuppression rather than ablation. The benefits of this last treatment modality for the various clinical indications range from cure to increased survival and stabilisation of the disease. HSCT presents a variable picture of toxicities related to the conditioning regime of extra-medullary manifestation, which can be potentiated by individual comorbidities such as mucositis, haemorrhagic cystitis, venocclusive liver disease, cardiotoxicity and others.

Complications related to the period of bone marrow aplasia include severe pancytopenia, which, if sustained, has a high mortality rate and requires support measures such as growth factors, haemotherapy replacement and antimicrobial coverage (febrile neutropenia and infections). The main graft-related complications are Graft versus Host Disease (GVHD) and, secondarily, graft loss. For each of these conditions, there are one or more support measures that will determine the final cost of the procedure (PATON, COUTINHO and VOLTARELLI, 2000).

The Brazilian Association for Organ Transplantation (Associação Brasileira de Transplante de Órgãos) reported that 1,813 bone marrow transplants were carried out in 2013, of which 1,144 were autologous and 669 were allogeneic throughout Brazil, according to the Brazilian Transplant Registry (Registro Brasileiro de Transplantes - RBT). These numbers have grown due to the campaigns carried out by the Ministry of Health. Since 2003, the number of donors has grown by 16,000 per cent. Before, the number of registrations was 35,000, today the Brazilian Bone Marrow Donor Registry has around 3 million registered donors according to the 2012 RBT. Despite the fact that Redome is the world's third largest register of voluntary bone marrow donors, the growth in the number of allogeneic transplants in Brazil remains slow and constant, at around 3% per year. Bone marrow transplants are complex and require a great deal of material resources, a greater number of nursing hours due to the long period of hospitalisation and the risks of the patient's condition worsening. The side-effects of the treatment, as well as complications, further increase the cost by extending the period of hospitalisation and increasing the consumption of medication.

This whole process, from finding a donor to finalising the treatment itself, generates a huge financial impact for the SUS of approximately R$14 million over a period of 100 days (approximate treatment time), given that the estimated cost per patient per day undergoing allogeneic transplantation is on average R$2,236.00. It is worth pointing out that 95% of all transplants are carried out in public institutions and are therefore paid for by the SUS (COUTINHO, 2009).

According to Nascimento et al (2012), the bone marrow transplant unit stands out for its use of a large number of high-cost technologies and nursing interventions, as the complexity of the treatment can lead to the patient needing intensive care, which is carried out in the transplant unit itself, which serves as an intensive care centre. A service with these characteristics requires a high level of nursing care, as oncology is a specialised service that requires more complex care. To this end, the Systematisation of Nursing Care is a valuable tool for structuring and organising nursing work.

According to the National Council of Health Secretaries (2003), the complexity of the health area in Brazil is mainly related to factors such as: multiple determinants of the population's state of health, requiring different types of actions and services to meet these needs; staff training and technological resources. As a result, the hospital service is becoming increasingly demanding due to the changes, having to adapt to the global environment, requiring a new organisation with the adherence of concepts that make the production process more customer-oriented and enable improvements in both productivity and financial levels.

In the area of care management, the SNC is a way of better operationalising the nursing process with the patient. It is a functional element, essential for nurses in optimising and managing nursing care, since it is based on technical and scientific knowledge, resulting in individual care with excellence (BACKES, SCHWARTZ et al,2005, p.186).

Although the SNC has been incorporated into the professional practice of some institutions, the difficulties in implementing it are related to the number of human resources available to carry out all the activities required by this methodology, which requires time to record and analyse data, and shows that nursing work processes are of great importance to the health institution and the SUS (TRUPPEL et al, 2009).

Amante et al (2009) state that the use of systematisation reduces the incidence and duration of hospital admissions and creates a cost-effectiveness plan, which is vitally important for institutions that perform haematopoietic stem cell transplants. Health resources are scarce and the reliability of data analyses can be increased by using processes that favour the planning, evaluation of care and control of hospital costs (PERES and ORTIZ, 2008).

According to the International Council of Nursing (1993), it is through knowing the costs of their interventions that nurses will generate positive changes, maintaining or generating a balance between quality, quantity and costs.

In the municipality of Rio de Janeiro, there are six institutions that carry out bone marrow transplants. They are: Hospital Universitário Clementino Fraga Filho (UFRJ), Hospital de Hematologia e Hemoterapia do Estado do Rio de Janeiro (HEMORIO), Hospital Universitário Pedro Ernesto (UERJ), Instituto Nacional de Câncer, Hospital São Vicente da Gávea, Hospital de Clínicas de Niterói, the latter two being private institutions.

According to Maiolino (2000), allogeneic haematopoietic stem cell transplantation is a highly complex procedure and a field that requires constant clinical studies that bring more and more scientific innovations, whether they are drugs and/or equipment. These scientific and innovative developments in the search for new options within treatment protocols are also accompanied by rising costs. It is necessary that, in parallel with scientific advances in inputs, there are also economic evaluation studies aimed at better allocating these resources.

Economic evaluation makes a comparative analysis between two or more competing alternatives, so it is possible to identify the values added to the technological alternatives and to the investment, making it a valuable management tool in the decision-making process for managers of healthcare institutions (NITA, 2010).

In view of the problem presented, the object of this study is the cost of haematopoietic stem cell transplants before and after the implementation of the Systematisation of Nursing Care (SNC), and the hypothesis to be tested is the reduction in the cost of allogeneic stem cell transplants caused by the SNC.

The hypothesis is based on the results of database searches on the subject of the study, which show that SNC improves the quality of nursing care and reduces the costs of the inputs used in care (AMANTE, ROSSETTO and SCHNEIDER, 2009; LEFREVE, 2005, FRANCISCO and CASTILHO, 2002).

General Objective:

To analyse the financial cost of allogeneic HSCT in a federal hospital in the city of Rio de Janeiro.

Specific Objectives:

To identify the cost items and sub-items related to HSCT in all its phases, as well as the time spent by the nursing professional caring for this patient before and after the SNC.

Evaluating the cost items and the nursing working hours in HSCT before and after the SNC.

To describe the cost-minimisation of allogeneic HSCT before and after the implementation of SAE.

1.2 JUSTIFICATION AND RELEVANCE

This study was justified by the proposals to expand the service at a haematopoietic stem cell transplant centre at a Federal Hospital in Rio de Janeiro, which aim to increase the number of transplant beds. Studies of this nature aim to contribute to the institutional management process and the development of care protocols.

Providing subsidies for the allocation of resources in public institutions is in line with actions to improve the management of the SUS, as recommended by the Ministry of Health (BRASIL, 2009).

For the aforementioned institution, which served as the field for data collection, the study can contribute by directly providing economic and technical data for its managers, with a view to improving quality, cost planning and auditing processes.

For nursing professionals, this work serves as a theoretical and practical basis for future research on this subject, since the conscious allocation of financial resources together with strategic planning and the execution of care tasks are necessary competences for the nurse of the 21st century (BRASIL. MINISTÉRIO DA SAÚDE, 2008).

The complexity of cancer patients and their treatment leads us to look for alternatives and solutions to everyday management problems in order to meet all the needs of the service, with care for the quality of the care provided, the logistics of human resources and material supplies.

CHAPTER 2

RELATED STUDIES

A review of the current literature on bone marrow transplant costs was carried out. Medline and LILACS databases were searched using the anagram PICO, which guides us to formulate the research question and search for the appropriate descriptors.

Table 1 - PICO anagram

P Patient	Bone marrow transplant patient
1 Intervention	Allogeneic transplantation
C comparator	1 nfection
The out comes	Costs

The next step was to search for descriptors for each anagram element in each database:

Chart 2 - Medline search

Descriptors for bone marrow transplant patients
Bone marrow transplant patients or in-patients
Descriptors for allogeneic transplantation
Allogeneic bone marrow transplantation or allogeneic haematopoietic stem cell transplantation
Descriptors for Infection
Infection or infections complications
Descriptors for costs
Costs and cost analysis or increased costs or costs or cost allocation or fees and charges

Chart 3 - Lilacs database search

Descriptors for bone marrow transplant patients
Bone marrow transplant patients or bone marrow transplant patients or haematopoietic stem cell transplant patients or hospitalised patients or hospitalised patients
Descriptors for allogeneic transplantation
Allogeneic haematopoietic stem cell transplantation or allogeneic haematopoietic stem cell transplantation or allogeneic bone marrow transplantation
Descriptors for cost
Costos or costs and cost analysis or costos y análisis de costo or costos directos de servicios or asignación de costos
Descriptors for infection
Infections

The search used the **Boolean operators "AND" and "OR" to associate** the descriptors. The inclusion criteria applied were: 1- articles published in the last 10 years; 2- English, Portuguese and Spanish languages; 3- articles available in full; 4- articles dealing with allogeneic bone marrow transplantation and costs in adults. The selection process consisted of choosing titles and then reading the abstracts. Once the studies had been selected, the articles were analysed in full. In order to organise and clarify the publications found, a table was used showing the name of the article, the database in which it was found, the year of publication, the authors, the journal and the methodology used in the article, as described below:

Table 4 - Articles found in the databases

Article name	Database	Year	Author	Journal	Methodology
Economic cost of peripheral blood progenitor cells in spain	Medline	2004	SANCHES-BLANCO, J. J. et al.	Barcelona Medicine	Partial cost analysis
Economic evaluation of intravenous itraconazole	Medline	2005	MOEREMANS, K. et al	Int j Haematol	Total economic evaluation
Economic evaluation of the treatment of systemic fungai	Medline	2005	COLOMBO, A. L. et al	Recent Programme Med	Total economic evaluation
Economic analysis of unrelated BMT: results	Medline	2005	LISSOVOY, G. et al	BMT	Total economic evaluation
Allogeneic transplant costs	Medline	2006	SVAHN, B. et al	Transplantation	Partial cost analysis
Treatment costs and survival in with grades III-IV	Medline	2006	SVAHN, B. et al	Transplantation	Partial cost analysis
Lower costs associated with HCT	Medline	2007	SAITO, A. M.et al	BMT	Cutting study
The clinical and pharmacoeconomic analysis of invasive aspergillosis	Medline	2008	CAGATAY, A. A. et al	Mycoses	Partial cost analysis
Cost of allogeneic hct with high dose regimens	Medline	2008	SAITO, A. M.	Biol blood marrow transplantation	Partial cost analysis
Costs of haematologic cell transplantation	Medline	2009	NAVNEET, S.M.	Ann HEMATO	Partial cost analysis
Economic evaluation of posaconazole vs fluconazole	Medline	2010	DE LA CAMARA, et al	BMT	Total economic evaluation
Evaluation of cvc coverings	Medline	2011	ANDRADE, A. M.	Paddle	Partial cost analysis
Predictors of mortality and cost in tcth	Medline	2012	KERBAUN, F.	Einstein	Partial cost analysis
Costs and cost-effectiveness of tcth	Medline	2012	PREUSSLER, J. M.	Biol blood marrow transp	Review study
High readmission rates are associated	Medline	2012	DIGNAN, F. L.	Clinicai transp	Partial cost analysis
Real-word costs of auto and allogeneic tcth	Medline	2012	BLOMMESTEIN, H. M.	Ann HEMATO	Partial cost analysis
Increased costs after allog haematopoietic	Medline	2012	SVAHN, B. M.	MO	Partial cost analysis

Cost-effectiveness analysis of voriconazole compared with fluconazole	Medline	2013	MAUSKOPF, J. et al	AM J. Health syst pharm	Total economic evaluation
Direct costs associated with febrile neutropenia in inpatients	Medline	2014	ZHOU, Y. P. et al	Cancer support care	Partial cost analysis
A cost and resource utilisation analysis of micafungin	Medline	2014	HEIMANN, S. M. et al	EUR J. Haematol	Partial cost analysis

After applying the inclusion criteria, 20 articles were selected, 18 in English and 2 in Portuguese. As can be seen in the table above, most of the articles were found in the Medline database and one in Lilacs. The year with the most publications was 2012, with 25% of the total.

According to the definition of economic evaluation, 14 articles presented partial cost analyses, 1 article was a review and 5 were full economic evaluations, in the case of the cost-effectiveness studies. Most of the professionals involved in the studies were doctors and one of the studies was carried out by a nurse. The approach was always aimed at comparing costs between different therapeutic modalities, sometimes accompanied by an analysis of the results in terms of survival.

The main objectives of these studies were: to characterise the costs of transplantation in the different therapeutic modalities, to identify the factors associated with increased or reduced costs, to calculate the real cost of treatment for reimbursement purposes and to analyse the economic impact of transplantation on the health system. In the surveys, many variables were found due to the different types of transplant, treatment protocols and sample sizes. All of this makes comparative studies in bone marrow transplantation difficult. Approaches to the costs of infections after transplantation have been well formulated, with cost-effectiveness analyses seeking to identify better alternatives for the use of antifungals, whether prophylactic or for treatment. Most of the studies whose methodology was full economic evaluation were related to the use of antifungals in febrile neutropenia during transplantation and shared the same objectives.

In their study, Moeremans et al (2005) aimed to compare the intravenous use of Itraconazole with conventional Amphotericin B. A decision tree was carried out which included the probabilities of response to treatment, toxicities, documentation of the microorganism and second-line treatment. The study showed that the use of Itraconazole in cases of neutropenia and fungal infection was much less costly in the cost-effectiveness analysis.

The study by Colombo et al. (2008) evaluated the use of intravenous Itraconazole empirically and prophylactically in cases of persistent febrile neutropenia. The use of this antifungal drug proved to be more cost-effective than other conventional antifungals, but more studies are needed to analyse which drug is most suitable for prophylaxis and empirical treatment during febrile neutropenia. In another study, MAUSKOPF(2013) sought to analyse the cost-effectiveness of Voriconazolevs Fluconazole, based on a randomised clinical trial. The main objective of this economic evaluation was to determine the cost-effectiveness of Voriconazole prophylaxis compared to Fluconazole prophylaxis based on the decision tree diagram, which concluded that the study shows that these drugs may be more cost-effective according to some subpopulations of patients undergoing allogeneic HSCT, maintaining the indication of both according to the underlying diseases. After reading the manuscripts and recognising so many variables, it is necessary to intensify economic evaluation studies for the treatment of transplant complications, especially fungal infections, due to their high costs.

This integrative review has shown us that there is still a need for specific studies relating the evaluation of hospital costs to bone marrow transplants. Considering that Brazil is a country of continental proportions, multicentre studies in different regions could provide a diagnosis of the transplant situation, the treatments carried out, as well as a database of complication rates, mortality and the use of financial resources for complications that could be avoided.

With the increase in life expectancy of the population and the increase in the complexity of care resulting from technological development, there has been an increase in the consumption of the nursing workforce, generating an impact on costs, especially in terms of the consumption of material resources (TELLES and CASTILLO, 2007).

It is well known that nursing services are often seen as a major expense generator in healthcare institutions, especially if this service is not presented in figures and data that can be analysed (BROKEL, 2008). Nurses, as those who know and are responsible for planning, coordinating, supervising and controlling their team's work, must be qualified to make decisions that ensure the rational use of available resources, without jeopardising the quality of care (SANTOS and CARVALHO, 2008).

In this way, professionals who know the costs of their care are able to understand the economic value of their work and help with planning and obtaining and maintaining resources. When nurses come up with alternatives that can add value to care practice, and when they add greater effectiveness, generating lower costs for the institution, they become more attractive to the needs of the market (MARGARIDO and CASTILHO, 2006).

CHAPTER 3

LITERATURE REVIEW

3.1 HAEMATOPOIETIC STEM CELL TRANSPLANTS

This transplant is a highly complex procedure and is the only treatment modality that offers a cure for some types of haematological neoplasms and disease-free survival (BOUZAS, 2000).

According to Nicolau (2004), haematopoietic stem cell transplantation is a therapeutic procedure in which these cells are infused intravenously, with the aim of restoring bone marrow destroyed by chemotherapy. It is a treatment option when haematological toxicity is the limiting factor or when the pathological process directly affects the bone marrow.

To arrive at this procedure as a therapeutic modality, scientific research had to travel a long way. Historical accounts say that bone marrow was used orally in 1891 by Brown-Sequard in an attempt to treat leukaemia. In 1937, Schretzenmayr administered bone marrow intramuscularly to treat infections with some positive results. In 1939, Osgood used bone marrow intravenously for the first time. In 1944, Bernard injected bone marrow into the medullary cavity without results. Lorentz, in 1952, showed haematopoietic recovery after bone marrow infusion in irradiated rats. George Mathé infused halogenated bone marrow intravenously into physicists accidentally exposed to irradiation with positive results. In 1961, McFarland identified the need for pre-infusion bone marrow conditioning. Later studies were carried out, including molecular genetics, where human leukocyte antigens (HLA) were identified, which made halogen transplantation possible, and George Mathé also identified GVHD (graft versus host disease) for the first time. Despite all these developments, bone marrow transplantation was universally instituted as a therapeutic modality in 1990 by Dr E. Donnall Thomas, who was awarded the Nobel Prize in Medicine for this achievement.

In Brazil, bone marrow transplantation began in 1979, in Paraná, by Dr Ricardo Pasquini, who was also a pioneer in unrelated and umbilical cord transplantation. The first transplant was carried out in Rio de Janeiro in 1982 by Dr Mary Flowers at Cemo/Inca. In São Paulo, the first transplant was carried out at the Sírio-Libanês hospital by Dr Dráuzio Varella (MACHADO, 2009).

In order to carry out a transplant, there needs to be a clinical indication, the existence of a compatible matched or unrelated donor (in the case of halogen transplants) and a transplant centre with adequate infrastructure in terms of logistics. According to Paton et al (2000), a large number of onco-hematological diseases, genetic disorders and immunodeficiency disorders have therapeutic indications for HSCT:

- **Onco-haematological:** Acute lymphoid leukaemia, chronic lymphoid leukaemia, multiple myeloma, Hodgkin's lymphoma, non-Hodgkin's lymphoma, follicular lymphoma, diffuse large cell lymphoma, mantle cell, acute myeloid leukaemia, chronic myeloid leukaemia, myelodysplastic syndrome and some solid tumours.

- **Non-malignant conditions**: aplastic anaemia (main indication), fanconi anaemia and sickle cell anaemia.
- **Genetic disorders:** osteopetrosis, Gaucher's disease and Hurler's syndrome.
- **Immunodeficiency disorders:** Wiskott-Aldrich syndrome.

Acute myeloid leukaemia accounts for 90% of acute leukaemia cases in adults, with an average age of 63. In the United States, 18,000 new cases of leukaemia are diagnosed each year, of which more than 12,000 are of the acute type (JEMAL, 2007). In Rio Grande do Sul, 100 new cases of AML are diagnosed each year at all ages, with an incidence of 0.5-1:100,000 inhabitants and an average age of 42 years.79% of new AML cases occur in adults (> 18 years) and, in five years, only 90 patients (17%) of the 532 patients diagnosed between 1996 and 2000 were alive (CAPRA, 2007). Estimates from the National Cancer Institute (INCA) put the number of new cases of leukaemia in Brazil in 2012 at 4,570 for men and 3,940 for women, 5 new cases per 100,000 men and 4 per 100,000 women (Brazil, INCA, 2011).

Currently, AML is the most frequent indication for haematopoietic stem cell transplantation (HSCT) worldwide. The benefits of allogeneic HSCT come from the destruction of malignant cells by the donor's immune system - the graft versus leukaemia effect (HOROWITZ, 1990 and VELARDI, 2012).

Transplants can be of three types: autologous transplant, when the haematopoietic stem cells are obtained from the patient themselves. It can come from bone marrow or peripheral blood, where the stem cells must be mobilised with the use of growth factors and/or prior chemotherapy. Peripheral blood is collected at a blood bank by apheresis and then cryopreserved. Autologous HSCT is indicated for patients with malignancies, the main indication being multiple myeloma. The sources of stem cells can be bone marrow, peripheral blood or umbilical cord. Bone marrow cells are collected in a surgical centre, where the donor is anaesthetised and the bone marrow is aspirated through multiple punctures in the ilium; umbilical cord cells are collected immediately after birth, with drainage of the cord by gravity and aspiration of the cord; peripheral blood cells are collected in a blood bank by apheresis through peripheral access or catheter (if the donor does not have viable peripheral access (MACHADO, 2009). The second type is allogeneic transplantation, when the bone marrow, peripheral blood or umbilical cord stem cells come from another person. Allogeneic transplantation can be related, when the donor is consanguineous, or unrelated, when the donor is not consanguineous. The third type is syngeneic transplantation, where the recipient is the donor's univitelline twin.

3.2 THE PROCESS OF HAEMATOPOIETIC STEM CELL TRANSPLANTATION

Transplantation can be divided into three phases. The pre-phase, which corresponds to the period from the indication, donor selection, recipient and donor examinations and assessments, as well as all the preparation for the procedure up to conditioning. The transplant itself, which is the day the cells are infused, and the post-phase, which begins on the first day after the new bone marrow is infused, including outpatient follow-up. This period will vary according to the patient (MACHADO, 2009).

The pre-transplant (outpatient) phase takes place as soon as the indication is confirmed and the donor is selected. Selecting a donor with an adequate degree of compatibility is one of the essential strategies for a successful haematopoietic stem cell transplant (HSCT). Among the genetic factors that have the greatest influence on the outcome of these transplants are the genes of the HLA system. Recognising the fundamental role played by the allogenicity of HLA molecules in post-transplant evolution has led to the development of new methodologies for identifying these classic transplant genes. Donors are included in Redome with medium resolution HLA A, B and DRB1 typing. Recipients should be registered in the National Registry of Bone Marrow Recipients (REREME) with high-resolution HLA class I and II typing because this information directs selection and thus speeds up the donor identification process (PEREIRA et al; SBTMO, 2012.). So far, the transplant centre's nursing staff has had no involvement. When the candidate arrives at the service, the pre-transplant assessment period begins with consultations with the multi-professional team, participation in educational talks, blood collection for laboratory tests, placement of a tunneled central venous catheter and donor preparation (clinical assessment) (MACHADO, 2009).

The pre-transplant phase has two stages: outpatient and inpatient. It begins when the patient is admitted to the unit, where the conditioning regime begins. It is the period in which the patient is subjected to high doses of chemotherapy and immunosuppression with the aim of promoting bone marrow aplasia, preparing the recipient not to reject the graft and eradicating residual disease. It covers the period from day -7 to -1 (minus seven to minus one). This conditioning includes hyperhydration, chemotherapy drugs that vary according to the conditioning regime, antiemetics, the use of intravenous uroprotectant and hydration if Bussulfan is part of the regime (ORTEGA, 2004). Preferably on the day before admission or on the day itself, the two-way tunneled central venous catheter is installed by a trained surgeon in the operating theatre. This catheter must be handled with strict aseptic technique and be handled exclusively by nurses.

There are different types of conditioning regimes that, since the beginning of transplantation as a therapeutic procedure, have emerged with the intention of reducing transplant-related mortality and expanding treatment to patients of older age and/or with comorbidities that would prevent it from being carried out. They are:

• **Myeloablative conditioning regimen (MA)**- is the combination of agents that should produce profound myeloablation and irreversible and fatal long-lasting pancytopenia in most cases, with the restoration of haematopoiesis through stem cell infusion.

• **Non-myeloablative conditioning regimen**: this regimen causes less prolonged cytopenia, with combinations and doses of chemotherapy that leave residual autologous bone marrow, which is eliminated with the infusion of donor stem cells, which initiates new haematopoiesis.

• **Reduced toxicity conditioning regimen** - this regimen conditions a reduction in toxicity, but with shorter-lasting pancytopenia associated with the infusion of donor stem cells, preferably obtained from peripheral blood, which reduces the duration of neutropenia. The reduction is around 30 per cent of the dose.

In the trans-transplant phase, which is the day the cells are infused, stem cells can be collected from bone marrow or peripheral blood, which requires more than two days. When the stem cell comes from another location, or from an unrelated donor, or from the umbilical cord, they are cryopreserved through a laboratory procedure by specialist biologists and biomedical specialists who
are frozen with DMSO (dimethylsulphoxide). Depending on the type of CTH bag, whether fresh or frozen, there will be specific preparation before infusion. A water bath with a thermometer, anti-histamine medication, antipyretics and/or corticosteroids, mannitol pre and post to provoke osmotic diuresis and thus eliminate the preservative from the bloodstream more quickly, and saline solution throughout the infusion procedure are used. All the resuscitation equipment must be close to the patient's room and the doctor must be present in the ward to deal with any eventualities. In the post-transplant phase, you start counting D+1, D+2, where in the conditioning protocol you must follow immunosuppression (cyclosporine or tacrolimus), metrotexate, and control of the symptoms consequent to the regime used. There are the results of chemotherapy as well as complications. The side effects of chemotherapy include mucositis, nausea and vomiting, diarrhoea, bleeding, alopecia and hyperpigmentation (FONSECA and SECOLI, 2008).

In the initial conditioning period, where a chemotherapy and/or radiotherapy regimen is implemented, the main chemotherapy agents most commonly used in halogen bone marrow transplantation are cyclophosphamide, fludarabine, bussulfan and metrotexate, The term myeloablation refers to the administration of total body irradiation (TBI) and/or alkylating agents at doses that do not allow autologous haematological recovery. Combinations of bussulfan and cyclophosphamide and TBI are considered myeloablative conditioning. Other agents have been introduced into the regimen at high doses, and in different combinations with cyclophosphamide or TBI, with the intention of further intensifying the conditioning, and these are melphalan, thiotepa, etoposide and dimethylbussulfan. These regimes are associated with high toxicity and mortality, associated with other factors such as the age of the patient, the underlying disease and the age of the donor. Over time, myeloablative conditioning has been used less. This reduction, although not very clear, may be associated with improvements in HLA-matching technology and better supportive care. We should be aware that complete myeloablation is impossible, which is demonstrated by the cases of autologous recovery with disease in many cases after transplantation. Other conditioning regimes have been developed with the aim of catering for the older population. The results with transplantation have shown a worsening of the toxicity and mortality levels related to the older population, where 50 years was the maximum limit allowed. In the non-myeloablative (NMA) regimen, cytopenia is minimal and toxicity is low, with immunosuppression. A good example is FLU-CY (fludarabine and cyclophosphamide), a combination developed in the United States. This type of conditioning requires peripheral blood stem cells, rich in T lymphocytes.

Examples of non-myeloablative conditioning regimes are fludarabine and cyclophosphamide (CY-

FLU), TBI 2 GY, TBI 1 GY, lymphoid radiation and anti-thymocyte globulin (ATG). In the trans-transplant phase, stem cells are collected, either from the bone marrow in the operating theatre or from peripheral blood in a blood bank. The stem cells can be infused fresh or cryo-preserved. The post-phase begins on the first day after cell infusion and it is during this phase that bone marrow aplasia will occur with the consequent inactivation of the immune system. Also in this phase, the patient shows the toxic effects of the conditioning regime. The first 30 days post-infusion, when bone marrow aplasia occurs, is the most critical phase of treatment (MACHADO, 2009).

After being discharged from the inpatient unit, the patient must continue to be monitored on an outpatient basis, in a day hospital, as in most cases there is a need for intravenous immunosuppressants until the patient switches to the oral route. Examinations are also still frequent, as the effects of the conditioning still last, as well as the need to monitor for infectious occurrences, bearing in mind that the patient's immune system will only partially recover after 100 days (ORTEGA, 2004).

3.3 STEM CELL TRANSPLANT COMPLICATIONS

HAEMATOPOIETIC

According to Paton et al (2000), complications can be related to the conditioning regime due to its toxicity, such as gastrointestinal, pulmonary, cardiac, hepatic, renal, neurological, infectious or other minor ones. Graft versus host disease is the consequent difference between histocompatibility antigens, causing a reaction between the marrow received and the recipient's organism. It can present itself in the period after the marrow infusion or later on. It is a desirable entity, as it is part of the expected cellular effect that contributes to a lower rate of relapses of the underlying disease, but on the other hand it can become extremely damaging to the point of causing irreversible damage to target organs (PINHEIRO, 2009).

Most of the side effects and complications occur in the first hundred days after HTC transplantation. Didactically, some authors categorise these effects into acute and late complications. The major complications result from the effects of conditioning regimes (chemotherapy and/or radiotherapy), loss of bone marrow function, reorganisation of the immune system - rejection and graft-versus-host disease. This phase is very critical and, as a consequence, the recipient is susceptible to a series of complications. These include

Gastrointestinal Toxicity

It is characterised by the occurrence of nausea and vomiting of varying intensity as a result of radiotherapy and chemotherapy, which are the first to be experienced by the patient. These side effects are managed with antiemetics, most commonly Ondasentrone (BMT Service, 2003, Paraná).

Oral mucositis is the most common complication after OMT, affecting around 90 per cent of patients. The regimens using bussulfan, total body irradiation and vepeside are the regimens most associated with mucositis. This is a debilitating and painful entity, with severe forms requiring the use of powerful intravenous analgesics and total parenteral nutrition. Mucositis is classified as mild, moderate and severe. The grades range from I to IV. In grade I, the mucosa is veiled with mild erythema; in grade II, the patient has erythema and moderate pain; in grade III, there are ulcerations and severe pain requiring systemic analgesia; and finally, in grade IV, there is bleeding from the oral mucosa. In addition to treatment with systemic analgesics and rigorous oral hygiene, laser applications are carried out by the dental service to accelerate the healing process. This complication can include infections in the lesions that require treatment with antifungal and antiviral drugs (CLIN J NURS, 2000). Another very common symptom is diarrhoea, which usually occurs in the first week after transplantation, also as a result of chemotherapy and radiotherapy. In this case, investigations are carried out to detect infections and/or acute GVHD.

Haematological complications

They occur due to bone marrow aplasia, which is only resolved by grafting the donor's haematopoietic stem cells. Marrow aplasia leads to a state of leucopenia and thrombocytopenia, requiring support measures to prevent bleeding and infections. Blood component replacements are frequent, both for red blood cells and platelets (ORTEGA, 2004).

Renal complications

They occur more frequently in the initial phase of transplantation, in around 30 per cent of patients. It results from the nephrotoxicity of total body irradiation and/or the chemotherapy drugs used during the conditioning period. Other factors that contribute to kidney failure are tumour lysis, dehydration and the use of other nephrotoxic medications such as Ciclosporin, Amphotericin B and Aminoglycosides. Ciclosporin serum levels should be monitored frequently to ensure effective therapeutic action and prevent side effects. Elevations in creatinine levels should be observed, directing drug readjustment. The use of haemodialysis may be necessary (HELAL, 2011).

Venocclusive Liver Disease (VOD)

It is a complication that affects between 5 and 55 per cent of transplant patients (PATON, 2000). It results from obstruction of the hepatic venules due to damage to the surrounding endothelial cells, sinusoids and hepatocytes. Clinically, VOD is characterised by jaundice, hepatomegaly, pain in the right hypochondrium, ascites and weight gain due to water retention, generally observed in the first few weeks after transplantation (Idem, 2000). The criteria for diagnosis are weight gain of more than 10 per cent of baseline weight, progressive jaundice and painful hepatomegaly (BEARMAN, 1995). This complication can lead to multiple organ failure and death.

Haemorrhagic Cystitis

It occurs in between 5 and 50 per cent of transplant patients and can appear at any stage of the transplant, be temporary or long-lasting. Transient and early onset cystitis occurs as a consequence of the action of cyclophosphamide metabolites, one of the most commonly used drugs in conditioning regimes. Acrolein, a cyclophosphamide metabolite that is toxic to the bladder mucosa, causes damage to the mucosa, leading to bleeding. Prophylaxis uses hyperhydration and a drug called Mitexan, which inactivates acrolein. Long-lasting haemorrhagic cystitis occurs due to viral infections such as Adenovirus and Polyomavirus (BK and JK). Severe haematuria occurs, associated with dysuria. This complication causes prolonged hospitalisation, suffering for the patient and increases the mortality rate (PATON, 2000).

Other complications

Dermatological toxicity, neurological, pulmonary and cardiac complications and infections. As infectious complications accompany all stages of transplantation, it is necessary to analyse them for a better understanding:

Infections

The types of infections, as well as their severity, differ from one patient to another, according to factors such as age, gender, underlying disease, conditioning regime and GVHD immunoprophylaxis. All patients undergoing HSCT face profound neutropenia (< 500 neutrophils) in the first few weeks after conditioning. The patient's immunity will only return to normal after about a year or, as in some cases, after years. Most of the infections that occur in the first 30 days after HSCT are related to bacteria and/or fungi. Infections are directly linked to the immunological changes caused by the transplant, the injuries caused by conditioning regimes and immunosuppressive therapy. The first infectious complications are bacterial, occurring during the aplasia period after conditioning. Between the second and third month after BMT, cytomegalovirus (CMV) infections are common. Effective treatment is with Ganciclovir. Medication to prevent and control infectious processes includes the use of antivirals (acyclovir, ganciclovir), sulfa drugs to prevent infection by pneumocystis carinii (pneumocystis jirovecii), antifungals (fluconazole, amphotericin B) and immunoglobulins. The service should have an infrastructure that favours the practice of infection prevention procedures (NUCCI, 2002).

Graft-versus-host disease (GVHD)

The term refers to the inflammatory reaction by the donor's immunocompetent cells against one or more of the recipient's organs, such as the skin, liver, lungs and gastrointestinal tract. It can have two distinct clinical presentations: acute, in which the donor's cytotoxic T lymphocytes attack the recipient's histocompatibility antigens, and chronic, in which immunocompetent lymphocytes differentiate in the recipient, involving, in addition to cytotoxicity, host immune dysfunction, allowing autoimmunity to develop (SOARES, 2007; PATON,2000).Among the reasons why stem cell transplant patients require intensive treatment are: pulmonary complications, drug toxicity, venocclusive disease, graft-versus-host disease, infections and dysfunction of other organs such as the kidneys (EMELE, 2009).

3.4 SYSTEMATISATION OF NURSING CARE AND THE NURSING PROCESS

The Systematisation of Nursing Care is defined as a functional element, essential for nurses in optimising and managing nursing care, since it is based on technical and scientific knowledge, resulting in individual care with excellence (BACKES; SCHWARTZ et al,2005, p.186).

Systematisation efficiently operationalises the Nursing Process in all its stages, which is defined as a work methodology based on the scientific method, addressing the dynamics of systematised and interrelated actions aimed at assisting human beings (HORTA, 1979).

The stages of the Nursing Process are history, diagnosis, care plan, prescription, evolution and prognosis. The nursing process is the expression of the clinical method in the profession and the SNC organises it into a system that guarantees its optimisation. All of these stages must be interrelated, seeking to adequately collect patient data (CARVALHO and BACHION, 2009).

The Federal Nursing Council (COFEN), through Professional Practice Law No. 7489 of 25 June 1986, establishes SNC as a private activity of nurses, the implementation of which is essential for differentiating and valuing nursing professionals. According to Pivotto and collaborators (2004), this systematisation, with a scientific basis, provides individualised and excellent care for the patient. As reported by Andrade and Vieira (2005). The implementation of the SNC guides the nurse's decision-making process in nursing team management situations. In addition, the SNC provides the opportunity for advances in the quality of care, which drives its adoption in healthcare institutions.

COFEN (2009), through Resolution 358/2009, determined that the implementation of the SNC should be carried out in all public or private environments where professional nursing care takes place.

According to Truppel et al. (2009), the SNC should be formally recorded in the patient's medical record and contain the nursing history, physical examination, nursing diagnosis, prescription and nursing evolution. Although each of these phases is called in different ways by different authors, they have the same concept.

Although the SAE is incorporated into the professional practice of some institutions, the difficulties in implementing it are related to the number of human resources available to carry out all the activities required by this methodology, which requires time to record and analyse the data (TRUPPEL et al., 2009).

Guimarães et al. (2002) state that it would be possible to obtain countless advantages by planning nursing care, such as directing their actions, making it easier to change shifts and, above all, an advantage for the patient by making nursing care personalised, efficient and effective. Through this process it is possible to achieve greater integration and interaction between the nurse and the patient, their family and the multidisciplinary team itself, increasing the quality of the care provided.

It is extremely important for nurses to carry out their actions based on technical, scientific and managerial knowledge, with the intention of preventing damage to patients' health and minimising hospital costs. However, we know that not all health institutions are able to obtain adequate resources to provide quality care, which is seen as an objective to be achieved, but which requires control in order to be able to evaluate the effectiveness of nursing actions. In order to analyse care in a given sector, measurable parameters need to be established, requiring the implementation and use of certain programmes and protocols.

From the above about haematopoietic stem cell transplantation, we can affirm the great importance of the SNC for nursing care provided in a transplant unit, given the patient's level of complexity (MAGALHÃES, 2005).

There are several nursing diagnoses identified in cancer patients, especially those undergoing HSCT. Examples include some of the most common, such as Risk of Damage to Body Skin Integrity, related to chemotherapy and radiotherapy, Risk of Infection, Abdominal Pain related to mucositis of the gastrointestinal tract, Hyperthermia related to altered immune response, Risk of Anaemia and Bleeding, Impaired Oral Mucosa and Impaired Nutrition. There are many nursing diagnoses that can emerge from the individual condition of oncological and bone marrow transplant patients. Care for the cancer patient begins on admission to the transplant centre, when the first assessment is made with a view to establishing nursing diagnoses and drawing up a care plan that meets the needs of that moment and prevents harm.

Care of the tunneled central catheter is one of the first procedures to be carried out by the nurse, with the intention of maintaining its permeability and preventing infections related to this type of venous device. Weighing the patient is very important in order to establish a standard before starting hyperhydration. Start complying with the conditioning protocol with the installation of hyperhydration using an infusion pump, prepare and administer anti-emetic medication, prophylactic intravenous sulfa and take blood for biochemical and haematological tests to assess the need for blood components. A rigorous water balance is necessary from the start of the protocol and most of the time it should be completed every four hours. As the conditioning progresses, the levels of dependence on nursing change, so that the constant presence of the team is necessary. Also very important, among other things, is the assessment of pain levels using the visual analogue scale, where the start of intravenous analgesia is determined. As the side effects of HSCT and complications arise, new diagnoses may emerge, leading nurses to constantly exercise clinical reasoning.

3.5 ECONOMIC EVALUATION

The World Health Organisation (WHO) states that approximately 75% of the Brazilian population depends exclusively on the Unified Health System (Sistema Único de Saúde - SUS), whose resources are much lower than in developed countries. Expenditure on health in Brazil in 2008, for example, was US 222.00 per inhabitant, while in the United States this figure exceeded US 5,000 (IBGE, 2008 and BRASIL, 2008).

In recent decades, the increase in average life expectancy, new technologies appearing on the market, the shortage of qualified labour and the lack of professional training in managing health units have led to a huge increase in spending. In this way, the search for efficient allocation of available financial resources is a growing concern among managers when deciding where to spend them. (BRASIL, 2008)

In this sense, studies on economic evaluation as a tool for managing and allocating resources are important for reducing costs for the health system and the client in the short, medium or long term (BRASIL, 2008).

The relevance of this type of economic analysis is based on evidence that comes mainly from two spheres. The economic sphere, which is based on the scarcity of resources in the face of needs, and the clinical and healthcare sphere, in which the incorporation of new technologies and the consequent increase in demand for goods and services demand more resources from the health sector (SANCHO and DAIN, 2008, p. 1279-1290).

The International Council of Nursing (INC, 1993) considered that some technologies are capable of reducing costs by increasing the efficiency and effectiveness of care, citing that some equipment can facilitate certain actions carried out by nurses, freeing them up for other activities. However, it recognises that these technologies are expensive.

In order to build an effective economic relationship, studies are needed in the health area to justify the incorporation of new technologies, always aiming to adjust costs and choose the best options for providing quality care and treatment. During their working day, both inside and outside healthcare institutions, nurses take part in procedures and interventions that use technological resources. Nursing therefore needs to be included in these studies and be able to evaluate the results of its actions based on costs.

In Brazil, studies of this kind are not common, perhaps because it is a relatively new resource or because of its complexity. As a result, some institutions use this tool only for fiscal purposes, failing to use it as a management resource, to the detriment of a more detailed evaluation that would allow them to carry out and maximise them efficiently (BRASIL, 2006, p.7-8).

Health economic evaluation is defined as the comparative analysis, in terms of costs and outcomes, of two or more competing alternatives. This allows us to identify the value added to new technologies and decide whether the value attributed to them justifies the proposed investment (NITA et.al, 2010).

Outcomes are the consequences resulting from the exposure of a group or individual to a causal factor. A positive health outcome is the main indicator of health benefit. It is therefore necessary to understand the meaning of the terms efficiency, effectiveness, efficacy and equity in health, which are used to contextualise outcomes.

According to the Ministry of Health (2008), efficiency is an economic concept that derives from the scarcity of resources and seeks to produce goods and services of interest to society at the lowest possible social cost. Effectiveness is the measure of the consequences or results resulting from the implementation of a health technology used in real or habitual situations of use. Efficacy is similar to effectiveness, except that the observation is carried out in ideal or experimental situations. Equity in health is the principle that ensures that resources are distributed according to the health needs of a given population (NITA et al, 2010).

A full economic evaluation compares the costs and outcomes of two or more health alternatives, otherwise we are dealing with a partial cost analysis (Idem, 2010).

Health economic evaluation studies include: cost-minimisation economic evaluation (CME); cost-effectiveness economic evaluation (CEA); cost-benefit economic evaluation (CBA); and cost-utilisation economic evaluation (CUE). Below are the main characteristics of health economic evaluation studies (NITA et al, 2010 and SOÁREZ, P. C. et al 2014).

The cost-minimisation economic evaluation (CME) compares the costs between alternatives whose outcomes are identical, seeking to choose the alternative with the lowest cost. The result is the total cost expressed in monetary units. For example, a study comparing dressing costs for pressure ulcer wounds, where two different dressings are used, but which have the same outcome, such as healing or reduction of the lesion. In the end, we will see which of the two options was less costly, which had a better cost-effectiveness ratio.

Economic cost-effectiveness evaluation (ECEA) is the difference between the costs expressed in monetary units of two or more health alternatives, divided by the difference between the effectiveness (clinical outcomes) of the alternatives to be compared, expressed in natural non-monetary units, such as years of life gained. For example, a study comparing the costs and effectiveness of topical therapy and compressive therapy for the treatment of vasculogenic ulcers. The costs of both alternatives in monetary units and the benefits (such as ulcer healing time) they provide will be evaluated, and the one with the shortest time will be chosen.

The economic cost-benefit assessment (CBA) identifies the costs and evaluates the benefits associated with different alternatives, expressed in monetary units. For example, a study evaluating the cost and benefits of nurse counselling for STD prevention. The costs of consultation for positive STD cases and the costs of counselling for the entire sexually active population are estimated, as are the benefits of this counselling, such as the cases of STDs prevented by counselling. The one with the best cost/benefit ratio will be chosen.

Cost-utility economic evaluation (CUE) is a type of cost-effectiveness in which the effects of an intervention are considered through health-related quality of life, such as life expectancy, years of survival, among others. Utility is a quantitative measure that assesses the client's preference for a given health condition. Generally, in this type of study the unit of clinical outcome is life expectancy adjusted for quality or quality-adjusted life years (AVAQ or QALYs). For example, as alternatives for a client with chronic kidney disease, treatment with weekly haemodialysis or kidney transplantation. The latter would represent a cure, but we can't ignore the probable rejection of the transplanted organ, which could lead to death. Therefore, not all patients prefer to take this risk and choose to live on haemodialysis, even though it represents a deterioration in quality of life. For this group of patients, kidney transplantation has an unsatisfactory cost-utility ratio (BRASIL, 2008).

In Brazil, studies on economic evaluation in health have yet to appear in large numbers and scope, although government agencies have been developing studies in this area in recent years (ARGENTA and MOREIRA, 2007; SCHULTZ et al, 2007, p. 358-64).

Although nursing is directly involved in the management process of the most diverse sectors of health institutions, studies of this nature are not its main modality. The ones that appear the most in nursing are those that develop partial cost analyses, a fact that challenges this area of health to improve its knowledge in order to contribute more efficiently to its sector or institution by being able to promote the allocation of available resources and improve their quality (MARGARIDO and CASTILHO, 2006, p. 427-33).

CHAPTER 4

MATERIAL AND METHOD

4.1 STUDY DESIGN

The method used was quantitative. It was an evaluative and retrospective study, which used economic evaluation through cost-minimisation to identify whether the cost of allogeneic transplantation was minimised after the implementation of Systematised Nursing Care (SNC). We opted for this type of design because we believed that the study's objectives could only be achieved through this economic analysis, which aims to compare costs between two alternatives with equivalent outcomes. In this study, we compared the costs of allogeneic HSCT before and after the implementation of SAE in an oncology hospital located in the city of Rio de Janeiro. Two time periods were analysed: one in 2006 when SAE did not exist and the other in 2012 when SAE had been implemented, as described below.

Cost-minimisation analysis is a particular type of cost-effectiveness study within economic evaluation and is used to analyse the costs involved in carrying out a particular intervention. In this type of analysis, the difference in costs between different alternatives that produce equivalent results is calculated (BRASIL, 2009 p.40).

Still according to the agency, (idem) the economic consequences of an intervention can be classified into three large groups: (1) direct costs, the cost of professionals, hospitals, inputs, medicines and other health-related costs that can be categorised as health costs, (2) indirect costs, those associated with lost productivity and (3) intangible costs, related to the intrinsic value of improved health status. This research only used direct costs.

The comparison groups must be homogeneous in terms of inclusion and exclusion criteria, the frequency with which the procedure is carried out, the duration of the treatment, the technique and the form of application must be identical, differing only in the costs of each strategy to be analysed (NITA, 2010, p.140).

This analysis helps health managers during the decision-making process regarding which technological alternative to invest in, and the one that offers the lowest cost is chosen (BRASIL, 2009 p.40).

4.2 LOCATION CHOSEN FOR THE STUDY

The study was carried out at the Bone Marrow Transplant Centre of a Federal Oncology Hospital in the city of Rio de Janeiro. This is a large public hospital with 14 inpatient beds and a day hospital, with capacity for approximately 16 patients. [a]The Bone Marrow Transplant Centre is located on the 7th floor of the hospital and is divided into an Inpatient Unit and an Outpatient Unit, comprising an outpatient clinic and a day hospital. The day hospital is open Monday to Monday from 7am to 7pm and the outpatient clinic from 8am to 5pm. The nursing staff includes 45 nurses and 30 nursing technicians. They work a 12x60 shift plus three additional shifts totalling 40 hours a week, as well as 10 day labourers from 7am to 4pm, Monday to Friday.

This institution performs an average of 200 transplants a year, including autologous, syngeneic, allogeneic related and unrelated transplants, using all sources of stem cells, umbilical cord, bone marrow and peripheral blood (RBT, 2012).

The institution was chosen because allogeneic transplantation is a highly complex procedure and the oncology hospital is one of the few that performs it in the state of Rio de Janeiro. Another factor that contributed to the choice of this location was that the researcher was part of the work team, making it easier to collect information.

4.3 POPULATION AND SAMPLING

The study's source population consisted of all adult patients who underwent HSCT in 2006 and 2012 and underwent all treatment at the study site.

All the medical records were analysed, totalling 158. The size of the population was justified by the following inclusion criteria: age range between 27 and 50, as this is the interval in which individuals are most productive and whose departure from the labour market has a major economic impact; the source of the transplant being bone marrow, as these are the most commonly used in HSCT, and the type of transplant being allogeneic related, as these are the most frequently performed in this transplant centre and do not depend on finding donors in bone marrow banks. Criteria were also applied based on the clinical conditions of the patients, to standardise the sample, also selecting those who presented the same patterns of complications/alterations, namely: same febrile neutropenia, similar infectious conditions, such as bloodstream infection, invasive fungal infection and pneumonia, the type of antibiotics and laboratory tests. After applying these criteria, the sample consisted of 14 medical records divided into two groups.

The first group consisted of seven medical records of patients who underwent allogeneic HSCT in 2006, when the unit had not yet implemented SAE. The second group consisted of seven medical records of patients who underwent HSCT in 2012, after the implementation of the ESS in the related sector.

The characteristics of the selected sample show that 57.1% of all transplant recipients had Acute Myeloid Leukaemia (AML) as their underlying disease, with an increase in 2012, which follows a worldwide trend, as shown in the table below (BALDOMERO, et al, 2011):

Table 5 - Transplant patients according to underlying disease - 2006 and 2012

Baseline diseases	Number			Percentage		
	2006	2012	Total	2006	2012	Total
Acute Myeloid Leukaemia	3	4	7	42,9%	57,1%	50%
Aplastic Anaemia	2	-	2	28,5%	-	14,3%
Baseline diseases	Number			Percentage		
Non-Hodgkin's Lymphoma	1	-	1	14,3%	-	7,1%
Leukaemia Acute Lymphocytic	1	1	2	14,3%	14,3%	14,3%
Leukaemia Chronic Myeloid	-	2	2	-	28,6%	14,3%
Hogdgkin's lymphoma	-	-	-	-	-	-
Syndrome	-	-	-	-	-	-

Myelodysplastic						
Total	7	7	14	100,0%	100,0%	100,0%

Allogeneic HSCT is the first line of curative treatment for AML in first remission (VISACRE, 2006).

The distribution in the table below shows that the largest number of patients who underwent transplants, both in 2006 and 2012, were between 27 and 50 years old.

Chart 6 - Transplant patients by age group - 2006 and 2012

Age group	Number			Percentage		
	2006	2012	Total	2006	2012	Total
27 to 34 years old	4	1	5	57,1%	14,3%	35,7%
35 to 42 years old	2	2	4	28,6%	28,6%	28,6%
43 to 50 years old	1	4	5	14,3%	57,1%	35,7%
Total	7	7	14	100,0%	100,0%	100,0%

For a long time, age was a limiting factor for transplantation. However, in recent years, the age limits for individuals to undergo HSCT have widened. For those who are still of working age, becoming ill with cancer limits them in their work activities and often leads to early retirement, with a huge social, economic, cultural, personal and emotional impact. From an economic point of view, this is of significant importance in developing countries, where economic instability is always present (SOUZA, 2010).We can see in the table below that, in terms of gender, there were no significant changes in the two years of the study:

Chart 7 - Transplant patients by gender - 2006 AND 2012

Gender	Number			Percentage		
	2006	2012	Total	2006	2012	Total
Female	3	4	7	42,9%	57,1%	50,0%
Male	4	3	7	57,1%	42,9%	50,0%
Total	7	7	14	100,0%	100,0%	100,0%

According to the Ministry of Health, the latest figures show that men are more frequent: 60 per cent worldwide and 54 per cent in Brazil (INCA, 2011 and IBMTR, 2010).

With regard to the source of the HSCT, despite the expansion of umbilical cord banks and cryopreservation techniques, bone marrow is still the preferred collection option for allogeneic transplantation, as it involves less donor time, as shown in the graph below:

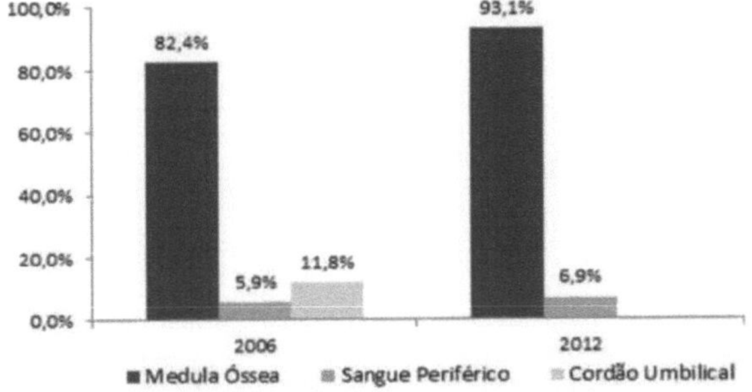

Graph 1 - Transplant patients according to HSCT source - 2006 and 2012

Of all the allogeneic HSCTs that were initially part of our sample (46), bone marrow was the predominant source of HTC for the two years studied, showing that it is still the most used source, especially in the case of related allogeneic HSCT. In other types of HSCT, other sources are used even more frequently, as in the case of autologous transplants. Matched transplantation is the type of allogeneic HSCT most used and has the highest incidence, constituting one of the inclusion factors for selecting the sample, as shown in the graph:

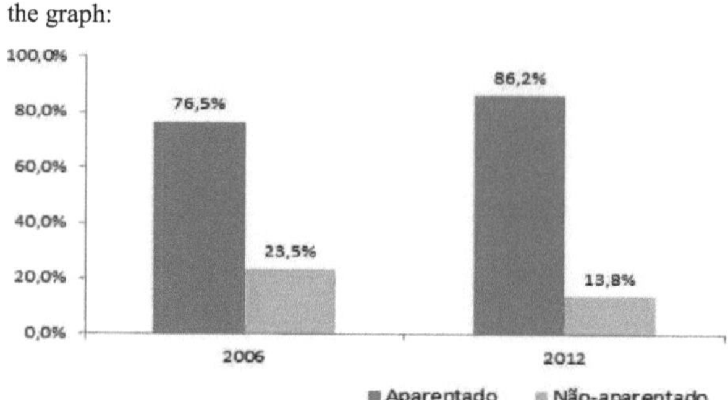

Graph 2 - Transplant patients by type of transplant - 2006 and 2012

This graph shows that the predominant type in the sample was related, in both years of the study.

One of the criteria for becoming a bone marrow transplant candidate is having a compatible donor. The chance of an individual having a compatible donor within their own family is 25 per cent, compared to 0.1 per cent in bone marrow donor banks. Obviously, patients who have a compatible family donor have a better chance of undergoing the procedure within the transplant queue (INCA, 2014). What we can conclude from analysing the characterisation of the participants in these two sample groups is that, in addition to the social economic impact that was not measured, these patients undergoing transplantation have a significant impact on the structure of the Unified Health System, as society loses out through unproductivity and loses out through prolonged, high-cost treatment. The economic deficit covers the private, institutional and/or government budget (DÓRO and PASQUINI, 2000).

All the patients in this sample are shown in the table below:

There are countless complications and side effects that affect transplant patients.

Chart 8 - Main complications in transplant patients

Complications	Description
Febrile neutropenia	Fever in the presence of neutropenia
Bacterial infection	Infection with microbiological and clinical documentation
Fungal infection	Infection with microbiological and clinical-radiological documentation

Coutinho 2009; Freifeld A.G.et al, 2010; Nucci, Maiolino, 2000.

These complications defined the uniformity of the sample in terms of the clinical standards of these patients and consequently the expenditure on the different cost items.

4.4 COLLECTING INFORMATION

The medical records of patients admitted to the transplant unit who underwent related allogeneic HSCT in 2006 and 2012 were analysed in order to gather data on transplant patient care.

The cost data was collected as follows: for both 2006 and 2012, the ABSOLUT hospital administration system was used, whose property rights have been acquired since 2000, and is integrated with Business

Intelligence (BI), which organises and cross-references patient data regarding diagnostic tests, medicines and supplies spent by the inpatient unit (CANCER FOUNDATION, 2011).

The data on pharmacy prices and warehouse materials is stored in this system, ABSOLUT. It was used to collect data on the prices of medicines and consumables. The prices of laboratory tests in 2012 were extracted from the same system, but the prices of tests in 2006 were extracted from the SUS SIGTAP table, because there was no data stored in the system for the years before 2008. With regard to antigenemia tests for CMV (cytomegalovirus), serum levels of CSA (cyclosporine) and Tacrolimus, and Galactomannan, the values were extracted from INCA's laboratories, and these values have remained the same since 2008. Searching for values for haemocomponents and stem cell bags also proved very difficult, as the institute has no records of the costs of concentrated red blood cells, platelets and stem cells.

In a cost survey carried out at the institute and the transplant centre, data was observed which took into account all the stages in the process of collecting, testing and finalising these components. The values of the blood components and the stem cell bags were taken from this study (COSTA, MOTTA and FARIAS, 2007).

The Ministry of Health's tenders, with INCA's code, were also used to extract data on materials and medicines that had not been found in the ABSOLUT system. In the search for price data, difficulties arose due to the lack of a database that concentrated all of the data.
items, both in the Bone Marrow Transplant service and in the National Cancer Institute in general.

In order to standardise the cost data relating to the different years with a six-year interval, we used the official US dollar monetary unit, which minimises the great instability of the national currency over this period of time.

For the data related to the value of the hour worked by the nursing professional, the institution's Human Resources Department (DRH) was consulted to acquire the Ary Frauzino Foundation (FAF) salary table. The salary scale for science and technology is available for public consultation on the Ministry's website. There are two different types of employment at the institution: employees hired under the CLT regime and those recruited by the Ministry of Health, Science and Technology. The average salary for each table was calculated and then an average between the two results to arrive at the value of the nurse's hour worked. The salary values used were the same for both 2006 and 2012, as there were no significant salary adjustments in this period.

Data was collected using an instrument designed for this purpose. It was an electronic form, developed in EXCEL, made up of four spreadsheets for collecting data on the pre-, trans- and post-HSCT phases and costs. Each of the three spreadsheets for the aforementioned transplant phases was subdivided into four parts:

- Quantities of inputs used;

- Quantity of medicines used;

- Type and number of tests carried out;

- Nursing professional - time used by the professional.

The fourth spreadsheet was used to collect data on the costs of medicines, tests, supplies and professional labour for each year (2006 and 2012), assigning the corresponding values to them, as this is a comparative study between these two different periods. Six months were spent collecting data from the medical records selected for 2006 and 2012, cataloguing the data in an electronic spreadsheet, searching for prices and statistical treatment.

4.5 DATA PROCESSING AND ANALYSIS

The data collected was organised into electronic databases by typing it into Microsoft Excel 2010 spreadsheets, from where it was exported and presented in graphs, charts and tables.

These spreadsheets gave rise to a database of transplants carried out in these two periods and the results were divided into three parts, the first referring to information on cost items, the second on costs and the third on cost-minimisation.

For the first part of the results, the following cost items were analysed: material inputs, medicines and tests carried out for the patient and the amount and time of nursing staff for each year of data collection (2006 and 2012) and for each phase of the transplant (pre-transplant phase; trans-transplant phase; post-transplant phase). The second part was divided into two subparts: the cost of the transplant in 2006 and the cost of the transplant in 2012, and the third part was the analysis of the cost-minimisation of the transplant carried out in 2006 (without implementation of the SNC) and the transplant carried out in 2012 (with the SNC already implemented in the Service).

The cost of the transplant was calculated from the costs found for each cost item (inputs, medicines, tests) plus the value of the time spent by the professional, calculated from the hour worked. By adding up the cost of the inputs, medicines and tests carried out and the cost of labour, we obtained the total cost of the procedure for each year studied. It is worth emphasising that only direct costs were used in this study.

Data collection was carried out with the help of a nursing student, after appropriate training. The student was trained in how to extract data from medical records, as follows: from the
paper spreadsheets (a copy of the electronic one) and in possession of the patient's medical records, we first analysed the medical prescriptions, copying the items of interest from them, then we read the patient's clinical evolution and finally we analysed the water balance. This triple check allowed us to be sure of what the patient had actually consumed during their hospitalisation. The data was collected jointly with the main researcher. This collaboration was necessary because the volume of data was very large and the time spent with each medical record was around 6 hours a day due to the complexity of the prescriptions.

The analysis of the first and second parts of the study, i.e. the data on inputs, medicines, exams and professional hours worked, as well as their costs, was carried out using Descriptive Statistics, which is used to describe and summarise the data so that conclusions can be drawn about them. Descriptive statistics refers to the way of presenting a set of data in tables and graphs, and a way of summarising the information contained in this data and some measures (FERREIRA, 2005).

In the third part, for the cost-minimisation (CM) analysis, the decision tree analysis model was used, using TreeAge Pro software from TreeAge Software Inc., version 2014.

Any decision analysis requires identifying the main strategies to be adopted, carefully predicting the likelihood of future events and the risks and benefits of each possible course of action. The Decision Analysis technique was developed to help healthcare professionals make rational decisions that reflect the best available evidence and the needs of the patient. The Decision Analysis method formalises the decision-making process, and for this we use the Decision Tree, which is a type of flow diagram that outlines the possible outcomes that could occur with each decision, and can calculate the probability and value of each decision (NITA et al., 2010).

The decision tree is the simplest of the decision analysis models and makes it possible to model a clinical situation or scenario that is not very complex. It is also useful when clinical problems are short-lived and do not recur (NITA, 2010).

We used the simple decision tree model because of the nature of the technology and intervention carried out by nursing staff (short and with uncomplicated outcomes). The decision tree is a graphic representation of how the possible choices can be related to the outcomes. Each possible choice is included in the analysis and represented as a branch. Outcomes that are not under the **decision-maker's control are represented by a "node"** and each decision alternative is associated with a numerical probability of the event occurring (HADDIX et al, 1996). The structure of a decision tree is explained in the table below:

Chart 9 - Structure of a decision tree

Type	Figure	Representation
Decision node	Square	Indicates a decision point between different alternatives; it is located at the beginning of the tree.
Probability node	Circle	Indicates a point where two or more alternative events are possible. These are shown as branches coming from a node.
Terminal node	Triangle	It indicates a point where a final outcome occurs.

Source: Nita, 2010.

We can consider a therapeutic strategy with three possible clinical outcomes: death, life without morbidity and life with morbidity. Considering that the values (utility) assigned by patients to each of these states is 0 for death, 1 for life without morbidity and 0.5 for life with morbidity, and assuming that the probability of each outcome would be 0.25, 0.50 and 0.25 respectively, we would have the weighted average for this therapeutic strategy 0.25 + 0.5 0.5 + 1.0 0.25 = 0.5. **This weighting is called the "expected value" of the strategy.** If we proceed in the same way for all the strategies or choices that make up the decision tree, **we can determine the one with the highest "expected value" and which will probably be the most desirable for** solving the problem in question.

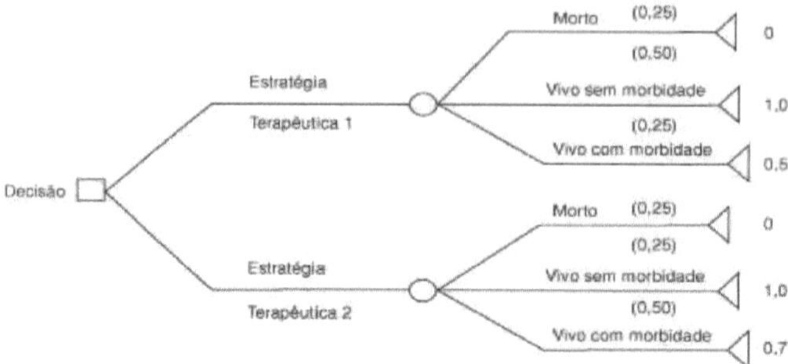

Figure 1 - Decision tree used to analyse health decisions

The square in the figure indicates a decision point: the circles indicate the chance points; and the triangles indicate the quantitative outcomes (measured in utilities). The numbers in brackets indicate the probability of each possible outcome occurring (CAMPOLINA and CICONELLI, 2006).

In this study, the total costs of related allogeneic HSCT carried out in 2006 and 2012 were compared to assess which had the lowest cost.

To compose the tree, it was necessary to identify the complications that generated additional costs. To do this, it was necessary to return to the collection site to identify them. After this identification, a consultation was held with specialists from the Bone Marrow Transplant Centre (doctors who are members of the bone marrow transplant service responsible for drawing up the protocols, with the intention of discussing the most relevant complications in this sample). The clinical complications defined in the tree were febrile neutropenia, bacterial infection and fungal infection. With the final consolidated costs for the two periods, it was possible to carry out an economic cost-minimisation analysis of HSCT at the time when SAE was not implemented and after its implementation. The alternatives analysed were placed in the first branches, HSCT 2006 and HSCT 2012. These branches were divided according to the main comorbidities: febrile neutropenia, bacterial infection and fungal infection.

The probability of each of these occurrences was determined by the average number of events found in studies whose objectives included the clinical characterisation of the client (VOLTARELLI. 2000; NUCCI,2000; HELAL 2011).The cost of the transplant in the two years studied came from the sum of the cost of material inputs, medicines, tests and the time spent by the nurse in providing care. Next, each comorbidity was subdivided into discharge and death, according to the outcome obtained with each one.The probability of each of these outcomes was obtained from the findings in the literature on studies carried out in other transplant centres (KARGAR, 2013; MENDES, 2010; COREY, 2006).To make up the decision nodes, probabilities were created:

Table 10 - Probabilities and cost variables

Name	Description	Value	Low	High
c_antib new scheme2012	-	1490.25	1192,2	1788,3
c_exams2006	-	9583.67	7666,936	11500,4
c_exams2012	-	5809.4	4647,52	6971,28
c_fungico12006	-	37488.0	29990,4	44985,6
c_fungico12012	-	14598.72	11678,98	17518,46
c_fungico 22006	-	72991.0	58392,8	87589,2

c_fungico 22012	-	10127.7	8102,16	12153,24
cjnsumos 2006	-	52513.9	42011,12	63016,68
cjnsumos 2012	-	66895.62	53516,5	93653,87
c_medicines 2006	-	198353.85	158683,1	238024,6
c_medicines 2012	-	53866.1	43092,88	64639,32
c_RH2006	-	324406.55	259525,2	389287,9
c_RH2012	-	305972.47	244778	367167
c_trat_bacteria	-	4211.76	3369,408	5054,112
p_high2006_antib	treatment. Neutral with line 1 atb and high 2006	0.8	0,64	0,96
p_0bito2006 _antib	treatment. Neut c/line 1 atb and obito 2006	0,2	0,16	0,24
p_alta2006novoesq	treatment. Neut w/ new esq atb and alta2006	0.6	0,48	0,72
p_obito2006novoesq	trat neut w/ new esq atb and obito 2006	0,4	0,32	0,48
p_altaantib2012	neutral treatment with line 1 atb and alta2012	0.8	0,64	0,96
p_obito_antib 2012	treatment. Neut with line 1 atb and obito 2012	0,2	0,16	0,24
p_altabact2006 alta	treatment inf bact c/ 2006	0.8	0,64	0,96
__ pobitobact2006	treat inf bact w/ obit 2006	0,2	0,16	0,24
p_altabact2012	treatment inf bact c/alta 2012	0.9	0,72	0,99
p_obitobact2012	treatment inf bact c/obit 2012	0,1	0,08	0,12
p_altafung 12006	antifungal treatment 1 w/high 2006	0.3	0,24	0,36
p_obito_fung1 2006	antifungal treatment 1 w/obito 2006	0,7	0,56	0,84
p_altafung12012	antifungal treatment 1 w/high 2012	0.6	0,48	0,72
p_obitofung12012	antifungal treatment 1 w/obito 2012	0,4	0,32	0,48
p_altafung22006	antifungal treatment 2 w/high 2006	0.2	0,16	0,24
p_obitofung22006	antifungal treatment 2 w/obito 2006	0,8	0,64	0,96
p_altafung22012	antifungal treatment 2 w/high 2012	0.4	0,32	0,48
p_obitofung22012	antifungal treatment w/obito 2012	0,6	0,48	0,72
p_altanovoesq 2012	trat neut novoesq c/alta2012	0.6	0,48	0,72
p_obitonovoesq 2012	trat neut novoesq c/obito2012	0,4	0,32	0,48
p_antbact2006	probab infection bact 2006	0.7	0,56	0,84
p_fungica2006	probab fungal infection 2006	0,3	0,24	0,36
p_antib2006	probab treat line 1 neut 2006	0.7	0,56	0,84
p_novoesq2006	probab trat novo esq 2006	0,3	0,24	0,36
p_antib2012	probab treat line 1 neut2012	0.7	0,56	0,84
p_novoesq2012	probab trat novo esq 2012	0,3	0,24	0,36

p_fungic12012	probab trat antifung 1 2012	0.8	0,64	0,96
p_fungic22012	probab trat antifung 2 2012	0,2	0,16	0,24
p_fungica2012	probab fungal infection 2012	0.1	0,08	0,12
p_bact2012	probab bacterial infection 2012	0,9	0,72	1,08
p_neutral2006	probab neutropenia 2006	0.45	0,36	0,54
p_neutral2012	probab neutropenia 2012	0,7	0,56	0,84
p_infection2006	probab infection 2006	0,55	0.44	0,66
pjnfection 2012	probab infection 2012	0,3	0,24	0,36
p_tratfung12006	probab antifungal treatment 1 2006	0.6	0,48	0,72
p_tratfung22006	probab trat antifung 2 2006	0,4	0,32	0,48

These variables represent the different scenarios encountered in bone marrow transplantation and make up the final cost of each alternative studied here.

The decision tree was divided into two main branches: HSCT in 2006 and HSCT in 2012. The next branches were the occurrence of febrile neutropenia and infection. According to the literature, neutropenia is defined as a decrease in circulating neutrophils. It can be mild, when the neutrophil count is between 1000 and 1500, moderate when the count is between 500 and 1000, and severe when the count is below 500 cells. In allogeneic bone marrow transplantation, neutropenia is of the severe type and is associated with immunosuppressive therapy, which makes the transplanted patient susceptible to febrile neutropenia and opportunistic infections, including sepsis (ZAFRANI and AZOULAY, 2014).

As a consequence of febrile neutropenia, there is an increase in overall costs, an impact on oncological prognosis and mortality (RABAGLIATI, BERTIN, CERON et al. 2014). Following the ramifications of the decision model used, each of these possibilities (neutropenia and infection) was subdivided according to the treatment carried out. The neutropenia branch was subdivided into first-line treatment and a new antibiotic regimen and the consequent possibilities of discharge and death. The infection branch was subdivided into bacterial infection and fungal infection, because according to the literature, these are frequent morbidities in transplant patients with persistent febrile neutropenia (NUCCI, 2000).

Next in the bacterial infection branch is antibiotic treatment (we're already on the third), resulting in two possibilities: death or discharge. The fungal infection branch was subdivided into two different antifungal treatments, one first-line and the other second-line, according to the literature, which are the most commonly used in clinical practice. The consequences of the respective treatments were discharge or death, with the results varying according to the probabilities found in the literature. It can be seen that different probabilities and expected values were collected for the two years in terms of costs. The analysis here looked for changes in costs after the implementation of the Systematisation of Nursing Care. The tree is represented in the figure below:

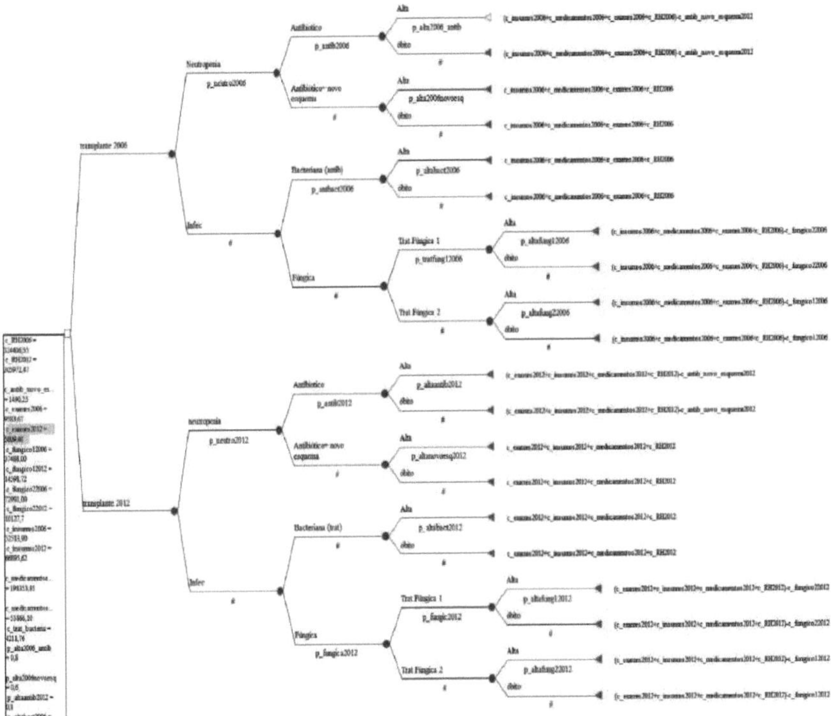

Figure 2 - Decision tree model used for this research

4.6 ETHICAL ISSUES

The study was approved by the Research Ethics Committee of the Federal University of the State of Rio de Janeiro, as the proposing institution, and by the Research Ethics Committee of the co-participating institution Instituto Nacional de Câncer, under research protocol CAAE 20869513600005285 and opinion 02/06/14 (Annex 1).

Resolution 466/12 of the National Health Council is based on the main international documents that have issued declarations and guidelines on research involving human beings, such as the Nuremberg Code (1947), the Declaration of Human Rights (1948), the Universal Declaration on the Human Genome, the International Declaration on Human Genetic Data and the Universal Declaration on Bioethics and Human Rights, as well as the provisions of the Brazilian Federal Constitution of 1988 and related Brazilian legislation, such as the Consumer Rights Code, Civil Code and Penal Code, Statute of the Child and Adolescent, Organic Health Law 8.080, of 19/09/90, among others. This Resolution incorporates, from the perspective of the individual and the collective, the four basic references of Bioethics: autonomy, non-maleficence, beneficence and justice, with the aim of ensuring the rights and duties that concern the scientific community, research subjects and the State.

Data from patients undergoing haematopoietic stem cell transplantation was collected directly from their medical records, which is why there was no need for a consent form, as no patient needed to be contacted. Due to its clinical complexity and the impossibility of accessing the participants, access to their data was justified to the Research Ethics Committee (REC) by means of a form justifying the absence of an ICF (Appendix 2), guaranteeing the confidentiality of the data collected and the non-disclosure of identities at any time, in accordance with CNS Resolution 466/12. There were some limitations during the course of the study, such as conflicting data in the medical records, for example, data on cell sources, whether related or unrelated, which took us a little longer to check during the research.Regarding the release of medical records for the study, despite the availability of the medical archive service, the institution was undergoing changes in the management of the archive, which is outsourced. This slowed down data collection, which involved a large volume of data to be processed.

CHAPTER 5

PRESENTATION, ANALYSIS AND DISCUSSION OF RESULTS

5.1 COST ITEMS AND SUB-ITEMS USED IN THE PRE, TRANS AND POST TCTH PHASES IN 2006 AND 2012

The study included 14 related allogeneic bone marrow transplant patients (7 in 2006 and 7 in 2012).

Having defined the number of patients and the characteristics of the medical records, we separated the transplant into three phases: the pre-, trans- and post-transplant phases, in order to identify the cost items, as shown in the tables below The number of cost sub-items (needles, equipment, catheters, among others, which were subdivided into cost items), as well as their diversity, led us to make groupings in order to make it easier to read and understand the items (FERREIRA, 2005).The table below shows the items and sub-items consumed in the pre-HSCT phase, characterised by chemotherapy conditioning:

Table 11 - pre-HSCT phase: Quantity of inputs, medicines and tests - 2006/2012

Pre-HSCT phase	Quantity 2006	Quantity 2012
Description		
Material inputs		
Needles (40x12, 30x8, insul)	1.073	996
Equipments (bi, doseflow, common)	497	376
Hygiene and dressings (gloves, gauze, adhesive tape)	1.732	532
Exam materials	128	134
Catheter	7	7
Syringes (60,20,10,05,03,01)	1.062	1.132
Total	**5.033**	**3.170**
Medicines		
Anti microbial	370	376
Intravenous solutions	1.856	1.780
Chemotherapy	1.151	3.358
Vasoactive drugs	0	8
Haemotherapy/blood components	111	128
Immunosuppressants and immunotherapies	128	50
Symptom control (anti-emetics, diuretics, anti-hypertensives, anxiolytics, analgesics, anticonvulsants, bronchodilators, anti-inflammatories, anti-allergics, gastric protectors, laxatives)	579	561
Antiseptic solutions and ointments	33	100
Other (folinic acid, trace elements, allopurinol, ursacol, vit b, vit k, multivitamins)	183	182
Total	**4.411**	**6.543**
Tests		
Biochemist	**700**	**748**
Haematological	**50**	**54**
Blood culture	**3**	**2**

This phase is characterised by prescriptions with chemotherapy protocols, where there are items common to all protocols, such as prophylaxis for pneumocystis carinii (now pneumocystis Jirovecii), with the use of intravenous Bactrim every 12 hours for 5 days. On day D-2, the other prophylaxes begin, after the previous one has been discontinued: antifungal and antiviral, with the use of Fluconazole and Acyclovir.

The consumption items that have an impact on this phase are those related to intravenous therapy. In the table above, we can see the quantities of the related consumption sub-items in 2006: syringes, needles and equipment totalling 2,632 units, all for infusions. In the hygiene and dressings sub-item, the consumption of gauze and gloves is also strongly related to venous therapy, as they are widely used in the handling of catheters and the preparation of medicines. Also in this sub-item grouping, the large quantity of gloves is due to protective isolation-related behaviours. In all phases this sub-item will be in high quantities for both 2006 and 2012. As far as the consumption item "medicines" is concerned, we're only going to see what happens in clinical practice, which is hyperhydration with volumes of up to 350ml/h in adults. Chemotherapy drugs are another important sub-item, as it is during this phase that they are administered, accompanied by the consumption of infusion pump equipment, the monetary value of which is high. It is important to emphasise the difference between the quantity of chemotherapy drugs in 2012. If we look at the epidemiological data presented above, we can see that the percentage of AML has increased, whose conditioning protocol is cyclophosphamide, and if the number of patients with the underlying disease AML has increased, consequently the consumption of cyclophosphamide has accompanied this change (BACIGALUPO and BALLEN, 2009).

In 2006, the consumption of immunotherapies was higher than in 2012, which can be explained by the presence in the sample of patients with the underlying disease AAS (severe aplastic anaemia), whose conditioning is done with anti-thymocyte globulin, which is an immunotherapeutic (MEDEIROS and PASQUINI, 2010).

In the other sub-item, the most commonly used medication was Ursacol, which is a liver protector and is used at all stages of the transplant, and Allopurinol, an important medication to avoid the tumour lysis crisis. As far as tests are concerned, biochemistry was the most commonly used due to the patient's predisposition to hydroelectrolytic imbalances at this stage. Electrolyte monitoring is necessary for electrolyte replacement purposes (TALLO, VENDRAME, LOPES et al, 2013).

The table below sets out the main nursing procedures in the pre-HSCT phase, where nurses work to minimise symptoms resulting from chemotherapy conditioning:

Chart 12 - Pre-phase: type of procedure performed by nurses according to their hours worked - 2006/2012

Pre-HSCT phase	2006	2012
Procedure	T=hour	
Blood collection	12,5	12,5
Catheter care	41,7	46,7
Preparation and administration of medicines	168	1243
Records	117	131
Total	339,2	1433,2

The procedures follow pre-established routines, such as the number of times routine blood is taken (always from 5am to 6am). There are extra collections, but they are requested with criteria, in case of extreme need. In the transplant department, we must remember the nature of the patients who are hospitalised: tunneled catheters, aseptic handling, interruptions to infusions in order to collect material for blood tests. For this reason, the criteria for requesting tests requires time-consuming nursing work and methodological rigour related to the prevention of catheter-related infections (SILVEIRA and GAL VÃO, 2005).

It can be seen that the time spent on records in 2006 was also significant, but it cannot be said that it was spent methodically, i.e. without systematisation. Nurses must manage their time by carrying out quality tasks (SPERANDIO and ÉVORA, 2008).

The number of nursing hours was higher in 2012 and, given the better results, we can say that Systematisation guided care in an organised and efficient way. The nurse's action must be one of care, identifying a need and knowing what to do to meet it (REPETO, 2003).

The trans-HSCT phase corresponds to the day of the bone marrow infusion, known in clinical practice as day zero or D0, when the bone marrow infusion takes place. This day is considered the rest day, because the conditioning chemotherapy has finished (ORTEGA, 2004). Let's look at the table below:

Table 13 - Trans phase - Cost items and sub-items: Inputs, medicines and tests - 2006/2012

Trans-HSCT phase	Quantity 2006	Quantity 2012
Description		
Material inputs		
Needles (40x12, 30x8, insul)	225	170
Equipments (bi, doseflow, common)	150	96
Hygiene and dressings	511	140
Test material	70	35
Syringes (60,20,10,05,03,01)	248	235
Total	1.430	683
Medicines		
Anti microbial	101	99
Symptom control (antiemetics, diuretics, antihypertensives, anxiolytics, painkillers), anticonvulsants, bronchodilators, anti inflammatories, antiallergics, gastric protectors, laxatives)	86	75
Antiseptic solutions and ointments	14	23
Vasoactive drugs (dobuta, nora, dopa, atropine, epinephrine, amiodarone)	0	1
Haemotherapy/blood components	15	10
Immunosuppressants and immunotherapies	18	20
Intravenous solutions	305	288
Other (folinic acid, trace elements, allopurinol, ursacol, vit b, vit k, multivitamins)	0	28
Total	539	544
Tests		
Biochemist	98	105
Haematological	7	7
Total	105	112
Grand total	2.074	1.339

Hyperhydration remains until the stem cell infusion, after which it is reduced. Consumption of items related to venous therapy is lower, as there are fewer drugs prescribed and, as already mentioned, most of the supplies are related to venous therapy. In this phase, there is greater consumption of gloves, gauze and other items related to patient care and handling, as well as catheter handling. With regard to medicines, the 2006 group spent more on medicines because the 2006 patients had a more complex clinical situation than the 2012 group. Analysing the databases of each patient in the group, we can see the need for patients to receive various electrolyte replacements, diuretics, antiallergic drugs and treatment of febrile neutropenia with more than one type of antibiotic. With regard to stem cell infusion, two patients had to receive pre-infusion medication, which included 20% mannitol, promethazine, paracetamol and hydrocortisone due to incompatibility between donor and recipient, in order to prevent transfusion reactions (PATON, 2000 and NEVES, 2008).

The 2012 patients had an uneventful infusion day with fewer drugs, and one patient used medication to prevent reactions before cell infusion.

The nursing procedures related to this phase are related to preparing the patient to receive the stem cells. On this day, the patient may experience hypertension, headache, chills, kidney failure, haemoglobinuria and decreased urine output. Nursing care aims to prevent, detect early and control these symptoms. The main ones are: Guiding the patient and family members on the infusion procedures, liaising with the cord bank professional (where the cells collected from any source are processed) to schedule the time for the HSC infusion, monitoring vital signs during the infusion every 15 minutes and in the second hour every 30 min, administering medications 30 min pre-infusion (antipyretic, antihistamine, osmolar diuretic, saline solution, corticosteroid). All these procedures are carried out as a priority, so that the patient is comfortable when the cell infusion begins (INCA 2008).

See the table below:

Chart 14 - Type of procedure performed by nurses according to their hours worked - 2006/2012

Trans-HSCT phase	2006	2012
Procedure	T=hour	
Cell harvesting	77	56
Blood collection	1,75	1,75
Catheter care	5,83	5,83
Preparation and administration of medicines	38	37,1
Records	14	16,3
Total	136,58	116,98

As mentioned above, on the day of the infusion, we have more time with the patient, which is justified by the cell infusion procedure that is carried out by the nurse, along with the care inherent to it such as monitoring, checking vital signs before, during and after, administering pre-infusion medication if necessary and/or afterwards in cases of reactions. Hygiene and comfort care such as bathing, dressing and activating the catheter must be carried out before cell infusion. This time requires dressing to prevent contamination of the cells (ORTEGA, 2004 and INCA, 2008).

In bone marrow aspiration, two nurses take part in the procedure, one as an instrumentator and the other responsible for homogenising the aspirated marrow and controlling the desired volume. During the procedure, the CTH are homogenised in a culture medium made up of saline and heparin, the purpose of which is to prevent the bone marrow from clotting. This procedure lasts an average of around four hours and these nurses are exclusively responsible for it. The aspiration and manipulation of the bone marrow containing the stem cells that will recompose the patient's marrow must be free from contamination, so it must be carried out in a protected environment as if it were a surgical procedure (MACHADO et al, 2009; CURCIOLI and CARVALHO, 2010).

The nursing time spent on this day in 2006 was higher. In the nursing activities on this day, the need for good management of the procedures to be carried out is very evident, such as forecasting the material needed for cell collection, confirming donor identification, blood type, confirming weight to calculate the volume to be aspirated, making donor/recipient identification labels to be used on the bone marrow bags and blood sample tubes, checking the autologous red blood cell bag for use on the donor if necessary. In short, a series of activities that must be coordinated according to the time available for the procedure, so that there are no unforeseen circumstances (FELIX, RODRIGUES and OLIVEIRA, 2009).

For so many tasks, the professional who takes part in this procedure, when it is not systematised, can lead to delays in carrying it out, rework and inversions in the stages of the flowchart, which can even cause problems during the transplant. Another difference was the time spent on nursing records, which was also higher in 2006. This year, nursing records were restricted to evolution, water balance and nursing prescriptions. As well as organising all these events, nurses in the transplant unit must set aside time to record their actions. Taking all the activities carried out in this phase together, without systematising these actions, we concluded that in 2006 there were a large number of hours worked by nurses that could have been reallocated to other activities. In hospitals, concern about costs and investments is mainly due to critical units, which demand a greater institutional burden, due to aspects related to material and human resources (VERSA, INOUE, NICOLA et al, 2011).

The implementation of Systematised Nursing Care, based on specific and critical knowledge of HSCT, is a fundamentally important tool for nurses to manage and optimise nursing care in an organised, safe, dynamic and competent way (BACKERS and SCHWARTZ, 2005).

The post-transplant phase is considered the most critical and is characterised by the manifestation of the toxic effects of conditioning. In this phase, the patient's clinical situation worsens and their feeling of discomfort is intense, especially in relation to the pain caused by painful mucositis, which requires intravenous analgesia (MAGALHÃES, 2005).

See the table below:

Table 15- Post-HSCT phase: Items and sub-items: supplies, medicines, tests - 2006/2012

Post-HSCT phase	Quantity 2006	Quantity 2012
Description		
Material inputs		
Needles (40x12, 30x8, insul)	3.762	2.780
Equipments (bi, doseflow, common)	2.067	1.783
Hygiene and dressings	3.600	1528
Test material	501	500
Syringes (60, 20,10,05,03,01)	3.630	3.413
Total	**15.131**	**10.004**
Medicines		
Anti microbial	2.151	2.805
Symptom control (antiemetics, diuretics, antihypertensives, anxiolytics, analgesics, anticonvulsants, bronchodilators, anti-inflammatories, antiallergics, gastric protectors, laxatives)	1.897	1.870
Antiseptic solutions and ointments	176	223
Vasoactive drugs (dobuta, dopa, nora, atropine, epinephrine, amiodarone	183	0
Haemotherapy/blood components	86	19
Immunosuppressants and immunotherapies	467	446

	2006	2012
Intravenous solutions	4.024	3.472
Others (folinic acid, trace elements, allopurinol, ursacol, vit b, vit k, multivitamins)	690	738
Total	**9.674**	**9.573**
Tests		
Biochemist	2.408	2.535
Haematological	172	169
Blood culture	32	39
Serious level	42	44
Total	**2.654**	**2.787**
Grand total	**27.459**	**22.364**

This is the most delicate phase of the transplant, as during this period the patient begins to show signs of toxic conditioning such as mucositis, diarrhoea and bone marrow aplasia (VOLTARELLI, 2000).

The patient becomes critical from a nursing point of view, with increased demand related to painful processes, gastrointestinal toxicity and pancytopenia, which makes them susceptible to infections, whether by bacterial translocation or external contamination (ZAFRANI and AZOULAY, 2014).

Protective isolation becomes essential at this stage. As a result, there is an increase in the consumption of supplies relating to catheter care, contact precautions, preparation and administration of antibiotics (COUTINHO, 2009).

This is the phase in which the most money is spent on antimicrobials and painkillers, as can be seen in the table above. The use of blood components is also present. If we compare the phases in terms of the quantities of inputs spent, we can see that there is a greater consumption of antimicrobials, medicines aimed at controlling symptoms, as well as intravenous solutions, but the latter are justified by the increased electrolyte replacement. It is worth noting that the post-phase studied here had a longer collection period than the others, lasting more or less four weeks, which is the average period of hospitalisation after cell infusion (CAPPELLANO et al, 2010).

Laboratory tests are an important component of this phase, as it is when serum dosages begin, the incidence of febrile episodes increases, and blood cultures are taken and biochemical dosages are also important (PATON, 2000).

Chart 16 - Type of nurse procedure according to hours worked:

Post-HSCT phase	2006	2012
Procedure	T=hour	
Blood collection	47,8	54.25
Catheter care	143,3	140,8
Preparation and administration of medicines	716,8	544,1
Records	401,3	394,3
Total	1309,2	1133.45

Procedures/hours worked: In this period of greater clinical compromise of the patient's general condition, nursing needs to be more present than in the other phases. The nurse's daily and continuous assessment can prevent adverse events, anticipate clinical behaviours and assist at the moment when the patient becomes more complex, even requiring intensive care. Although we are talking about a longer period of time, we are also referring to the level of complexity the patient reaches. Often a patient remains in critical condition for weeks or months, with high demands on nursing staff. The family, who are always present as carers, play an important role and it is at this stage that the nurse's interaction with the family is most necessary. The process of educating patients and their families and preparing them for discharge and/or advising nurses on the role of carers in relation to their critically ill relatives are tasks that require nursing time and methodology (MAGALHÃES, 2005).

As the data has shown, the patients in 2006 were more complex, requiring more nursing time to carry out their actions, while in 2012, the patients had fewer complications, especially those of an infectious origin, consuming less of the nurse's time (TELLES and CASTILHO, 2007).

5.2 VALUING COST ITEMS

Below is a consolidated summary of all the costs of related allogeneic transplants carried out in 2006, when SAE had not been implemented:

Table 17 - Total cost of bone marrow transplantation in the pre-, trans- and post- phases - year 2006

Phase	COST ITEM	Time in minutes of the enf.	Qty of items	Total in R$
Pre-HSCT phase	A - inputs	-	5.033	R$12.506,62
	B - medicines	-	4.411	R$ 29.060,95
	C - exams	-	752	R$1.695,98
	D - human resources	20.320	-	R$61.558,02
	Total	**20.320**	**10.196**	**R$104.821,57**
Trans-HSCT phase	A - inputs	-	1.430	R$2.850,88
	B - medicines	-	539	R$5.768,52
	C - exams	-	105	R$ 234,22
	D - human resources	8.205	-	R$ 24.856,48
	Total	**8.205**	**2.074**	**R$ 33.710,10**
Post-HSCT phase	A - inputs	-	15.131	R$ 37.156,92
	B - medicines	-	9.674	R$163.546,40
	C - exams	-	2.654	R$7.653,47
	D - human resources	78.560	-	R$ 237.992,05
	Total	**78.560**	**27.459**	**R$ 446.348,84**
Grand total	-	107.085	39.729	R$ 584.880,51

Valuing costs involves two main variables: the quantity of human resources and materials needed, and the unit value of this resource (ANNEX 3) must represent the opportunity cost of the resource. Opportunity cost is economic efficiency and is obtained through the maximum possible output from a given amount of resources (NITA, 2010).

For 2006, we'll discuss one item at a time, in the order in which they appear, starting with inputs. The total cost of inputs in 2006 was R 52,514.42, totalling all the phases. In the pre-conditioning phase, 5,033 items were consumed. This amount corresponds to the daily routine of hygiene and comfort, catheter care and medication administration (MAGALHÃES; MATZENBACHER and PACHECO, 2005; ORTEGA, 2004).

During this period, we can say that the patient's clinical condition is better, because they have been prepared and assessed prior to their hospitalisation, only to be subjected to aggressive treatment afterwards. The patient's performance must be good to withstand all the conditioning with minimal development of morbidities. This prior assessment is very important and has a prognostic factor (KERBAU et al, 2012).

The supplies used in the bone marrow transplant centre are often very expensive. In the pre-transplant phase, the most expensive consumable sub-item is the tunneled central catheter, valued at R$400.00 in 2006. Each patient receives one of these catheters before being hospitalised (SILVEIRA and GALVÃO, 2005).

In second place, the most expensive sub-item is infusion pump equipment, worth R$64.34 each. This equipment is used for venous hydration, chemotherapy and cyclosporine. They are only used once for chemotherapy infusions, so if you have two chemotherapies to infuse, two different sets of equipment are used. The hydration circuits are changed every day, as is the cyclosporine, which in 2006 was administered in continuous 24-hour infusions.

If we add up the cost of this sub-item per day, the patient consumes around 15 pieces of equipment per week in the pre-transplant phase alone, which would give a value of R$965.10 per patient in the first week of hospitalisation. The third most expensive sub-item is doseflow equipment, costing R$18.50, which is flow-controlled equipment with a volume device in ml/hour. This equipment is used for electrolyte replacement. The 2006 group of patients received around 13 electrolyte replacements using the *"doeeflow"* equipment, at a cost of R$240.50. In oncology, due to the risk of adverse effects from the solutions used for treatment, the industry launches products with technological innovation aimed at minimising errors (COREN-SP, 2010).

In the trans phase, the sub-item with the highest individual cost is still infusion pump equipment and *"OoeeOov"*. There is one item that at first seems to have a low individual cost, which is the box of procedural gloves, the value of which in 2006 was R$21.00. However, the consumption of this material is high and ends up adding to the total costs. Gloves are given to the carer to handle excreta and to sanitise the carer. Nursing staff also use a lot of gloves when handling medication circuits and for sanitising. If a patient uses 4 boxes a week, this material alone adds up to R$84.00. In the post-transplant phase we have a greater quantity of materials due not only to the 4-week period, but also to the fact that this is the period of greatest clinical compromise for the transplanted patient (COUTINHO, 2009). During this phase, a greater number of infusion pump sets are consumed due to the intravenous solutions for analgesia, hydration, cyclosporine, some antifungals and immunoglobulin, if any (ORTEGA, 2004; INCA, 2008). Some patients consume 5 to 6 infusion pumps. The cost is very high when the patient's general condition is compromised. On average, a stable patient spends three BI sets a day at this stage, totalling R$1,351.14 in a week. As nursing demand increases with bed baths, dressings and repeated intimate hygiene, glove consumption can rise considerably (LACERDA, LIMA and BARBOSA, 2007).

With regard to the item "Medicines", in the pre-phase, what had the greatest impact were the costs of chemotherapy, which totalled around R$23,000.00, i.e. almost all the costs of medicines in this phase.

In the trans-phase, drug consumption was lower, considering that it corresponds to a single day. In this group, only two patients required pre-bone marrow infusion medication. Two patients were already using antibiotics. Immunosuppression was present in all patients. The cost of cyclosporine in 2006 was R$60.15 per ampoule. The total cost for the seven patients was R$661.65. The post-transplant phase is where the most money is spent on medication. In 2006, the total cost of medication in this phase was R$163,546.40. The most expensive sub-item was the antifungal Caspofungin, which cost R$1,400.00 a bottle in 2006. The second most expensive drug was also an antifungal, Amphotericin B, a lipid complex that cost R$1,105.00 per vial (BROWN, 2010; DYKEWICZ, 2001; IRWIN; KLEMP; GLENNONC et al, 2011).

Cyclosporine was maintained for the entire period for all patients, and a fourth-generation cephalosporin, beta lactam, was also used, whose spectrum is broad for various strains, both Gram negative and Gram positive. This year, the cost of this drug was R$16.5. R$110,479.82 was spent on antifungal treatment alone. The impact of infectious complications is great on transplant costs (PAGANO et al, 2011 and RABAGLIATI, et al, 2014).

Working to protect the patient from the factors inherent in the clinical situation caused by the treatment itself has always been a major challenge for transplant centre nurses, and we bear a huge burden of responsibility, since we are responsible for managing protective isolation and using most of these materials (LACERDA, LIMA and BARBOSA, 2007).

The consumption of vasoactive drugs was significantly high, once again indicating the severity of the patients' condition. In the event of haemodynamic instability, drugs are administered to maintain acceptable blood pressure levels, cardiac function and, consequently, renal perfusion. These drugs are administered in the form of *"deeeping"* via an infusion pump. In this case, the patient's monitoring is complete: cardiac, pressure, oximetry and MAP (mean arterial pressure). This is the intensive care transplant patient, whose care is carried out in the transplant centre itself with the same nursing team. This is where we understand the large number of nursing hours consumed. The mortality rates for patients who reach this clinical situation are high (PAGANO et al, 2011).

With regard to tests, which is the third item, routine tests (daily blood count and biochemistry) were the norm in 2006. Total expenditure on tests this year was R$9,583.67. Less was spent on microbiological screening, as some types of tests were not available at the time, such as Galactomannan. Blood cultures were taken in cases of febrile episodes.

In 2006 there was no precise methodology to guide the application of the nursing process and we didn't work with diagnoses according to the Systematisation. We can see, then, that nursing work loses efficiency when carried out without proper management of its actions (TRUPPEL, 2009).

The fourth cost component is human resources, which had the biggest impact on the total cost for this year, totalling R$324,406.55. Analysing the pre-HSCT phase first, we see that in 2006 R$61,558.02 was spent on assisting patients undergoing conditioning. In this phase, the patient's reactions to the chemotherapy infusion take place, which consumes the nurse's time in attending to these symptoms with the intention of controlling them. Also, the preparation and administration of chemotherapy drugs takes up time that could be used for better care and support for the patient and their family. At the time, all chemotherapy was prepared in the unit. During the day it was prepared by the nurse in charge of dilutions and at night by the nurse assistant. The department always had a laminar flow cabinet and all the necessary PPE. All this time could be better distributed to improve the quality of care provided to transplant patients. In the trans-phase, the total cost was R$24,856.48, which includes care on the day of the bone marrow infusion. The lack of systematisation leads to rework and under-utilisation of professionals, as well as wasting resources on labour that could be spent on hygiene and comfort, clinical assessment of the patient and infection prevention measures (CASTILHO, 2002; GARBIN, 2011).

In the post-transplant phase, the cost of labour was around R$237,992.05, justified by the complexity of the patient in this phase and the evidence provided by reading the medical records that the patients transplanted this year had reached a degree of severity compatible with an intensive care unit.

The table below shows a consolidated summary of all the costs of related allogeneic transplants carried out in 2012, following the implementation of SAE:

Table 18 - Total cost of bone marrow transplants in the pre-, trans- and post- phases 2012

Phase	Cost group	Time in minutes	Qty of items	Total in R$
Pre HSCT	A - inputs	-	3.170	R$ 10.653,07
	B - medicines		6.543	R$7.380,83
	C - exams		803	R$ 644,76
	D - human resources	25.960		R$ 78.644,01
	Total	**25.960**	**10.516**	**R$ 97.322,68**
Trans-phase HSCT	A - inputs		683	R$4.109,71
	B - medicines		544	R$4.239,87
	C - exams		112	R$105,98
	D - human resources	7.025		R$21.281,75
	Total	**7.025**	**1.339**	**R$ 29.737,31**
Post phase HSCT	A - inputs		10.004	R$ 52.132,84
	B - medicines		9.573	R$ 42.245,40
	C - exams		2.787	R$5.058,66
	D - human resources	68.015		R$ 206.046,71
	Total	**68.015**	**22.364**	**R$ 305.483,60**
Grand total		**101.000**	**34.219**	**R$ 432.543,59**

In 2012, the total cost of inputs was R$66,895.62. This was an important cost in terms of monetary values, and it is interesting to note and follow the outcomes that followed during this year with the sample in question. Lower infection rates, less complexity and worsening of patients as a result of infectious complications (BROWN, 2010).

In the pre-phase, the most expensive input sub-item was the infusion pump with a monetary value of R$90.99. In 2012, consumption of this equipment increased because the intravenous immunosuppressant, cyclosporine, began to be dosed twice, every 12 hours, and two pieces of equipment were used.

As immunosuppression only begins two days before the infusion, each patient spent four pieces of equipment, totalling R$1,273.86 (ORTEGA, 2004).

The second most expensive sub-item this year in terms of inputs was doseflow, which cost R$15.50 per unit. Patients in this group received five electrolyte replacements in the pre-phase, totalling R$77.50.

There was expenditure on lower-cost items, which when consumed in large quantities end up adding to total costs, such as catheter care and the difference in expenditure on infection prevention and control actions determined by the nursing diagnosis of risk of infection (SILVEIRA and GALVÃO, 2005).

This was a fundamental factor in the development of the nursing care given to the patient that year, which was effective in the outcomes that followed, namely the resolution of febrile neutropenia and infection (MAGALHÃES, MATZENBACHER and PACHECO, 2005).

Protective isolation has been recommended for stem cell transplants for many years, but other preventive measures are important in caring for this patient, such as microbiological screening by the nurse on admission, using a nasal swab, instituting strict contact precautions with supervision of the procedures for patients with antibiotic-resistant germs and advising family members on the importance of adhering to them. These procedures require a higher cost in terms of procedural gloves, gauze, chlorhexidine degermant pads and non-disposable cloaks (CANTONI et al, 2009).

In the trans-phase, the input sub-items with the highest values were the same as in the pre-phase. In the post-phase, the input sub-items that had the greatest impact were the equipment from the other phases and the hygiene kit, an imported material that comes with a cleaning product and can be used for both bathing and intimate hygiene. If each patient consumes one of these kits, it will cost the seven patients R$324.80.

A total of R$53,866.40 was spent on medicines. In the pre-phase, we consumed a lot of oral chemotherapy (Bussulfan tablet). This chemotherapy is part of the conditioning protocols. In 2012, Bussulfan was administered orally because it was not available intravenously in Brazil. 16 doses are administered, with a large number of tablets (around +/- 35 per dose). The total cost of Bussulfan was R$96.17. The administration of bussulfan is accompanied by anticonvulsants (intravenous diazepan), which is a prophylaxis of seizures during conditioning with this type of chemotherapy. Also in the pre-phase, we have a large number of symptom controllers such as anti-emetics, diuretics and anti-hypertensives. In the trans phase, we have the use of drugs that prevent transfusion reactions, such as hydrocortisone. There was a total of only three replacement haemocomponents, the rest being bone marrow stem cells. The cost of haemotherapy was R$ 225.30. In the post-phase, we observed a significant amount of broad-spectrum antibiotics from the cephalosporin class, which are widely used as a first-line treatment for febrile neutropenia and are inexpensive. The cost of antibiotic therapy was R$7,809.74.

There was no need for amines, which means that the level of complexity of the patients was low, with a reduced demand on nursing from the point of view of the degree of dependence (MAGALHÃES, MATZENBACHE and PACHECO, 2005).

In terms of tests, the sub-item that had the biggest impact on this sample was the serum cyclosporine level, which cost R$33.00. The level is collected from peripheral access and is taken three times a week. The total cost of serum cyclosporine dosages was R$1,683.00. Galactomannan is used to screen the patient for fungi. It is only carried out when requested by the infectologist. Its cost was R$ 30.00 (PAGANO et al, 2011).

The fourth cost component is human resources, and these are worthy of note in this essay, due to the large number of hours spent assisting transplant patients. Total labour costs were R$305,483.60. How were these hours distributed in the care? In the pre-transplant phase, R$78,644.01 was spent on assisting the patient undergoing conditioning; in the post-transplant phase, R$21,281.75 was spent, corresponding to the day of infusion and the corresponding assistance; and in the post-transplant phase, R$206,046.71 was spent, corresponding to assisting the patient in their most vulnerable phase of the transplant (ORTEGA, 2004).

This year, the SNC had already been implemented and the effects of this SNC-directed care could be seen very clearly in this study: the outcomes related to infections, morbidity and mortality were much lower and satisfactory for what is expected of a bone marrow transplant centre. Nurses must use critical thinking skills to make decisions, as they are the team leaders and systematise care, directly influencing the results of healthcare (FELIX, RODRIGUES and OLIVEIRA, 2009; SMELTZER and BARE, 1998).

When the nurse-patient ratio is low, it improves health care results (ROTHEBERG, ABRAHAM and LINDENAUER, 2005). However, care needs to be optimised with a work methodology, otherwise we could make the mistake of having a large number of nurses without effective care. Some studies report that the experience of nurses can have an impact on hospital costs, but few studies address this relationship (THUNGIAROENKUL, CUMMINGS and EMBLETON, 2007).

SAE improves communication between team members and this is reflected in the prevention of adverse events and contributes to patient safety, which implies better overall care results (PORTAL and MAGALHÃES, 2008).

Faced with a result that shows a reduction in costs with effective care, we must conclude that this data constitutes important proof, showing that SAE is definitely a management tool that should be considered essential for reducing hospital costs and, in particular, for reducing treatment costs in bone marrow transplantation.

Nursing professionals are the biggest providers of care and represent the largest portion of the human resources workforce, being responsible for a large part of the final product of care (SILVA, JODAS, BAGGIO et al, 2012).

Here is a summary of the incidence of complications according to the literature:

Chart 19 - Complications associated with HSCT

Complications	Occurrence rate
Febrile neutropenia	45%
Bacterial infection	55%
Fungal infection	20%

Source:Nucci,2000; Penack, O. et al 2014; Kargar, M. et al,2013.

Here is a summary of the costs according to the infectious complication for the two years:

Table 20 - Cost of treatment for HSCT complications

Complication Description	Pre-HSCT phase 2006	Pre-HSCT phase 2012	Trans-HSCT phase 2006	Trans-HSCT phase 2012	Post-HSCT phase 2006	Post-HSCT phase 2012
Neutropenia c/ treatment 1	R$2085,36	R$378,9	R$453,34	R$50,52	R$13328,19	R$5279,34
Neutropenia w/treatment 2	R$28,118	-	R$52,236	-	R$1490,254	R$897,84
Bacterial Infection	-	-	-	-	R$4211,76	R$1632,56
Fungal infection treatment 1	-	-	-	-	R$37488,00	-
Fungal infection treated 2	-	-	-	-	R$72991,82	-
Total	R$2.113,482	R$378,90	R$505,576	R$50,52	R$110,479,82	R$7.809,74

5.3 COMPARING COSTS BETWEEN 2006 AND 2012

By comparing the costs between 2006 and 2012, we can see from the survey of direct expenses that there was a higher consumption of items in the first year. The

The total cost of medication was R$198,353.92 in 2006 and R$53,866.10 in 2012. There was a particular difference in the post-HSCT phase, where the cost of medication was much higher. There was an influence from the individual prices of some cost sub-items related to medicines, such as the monetary value of a particular antifungal used in 2006, which was much higher than it is today. The fall in drug prices is often due to the fall in patents. Most medicines come from other countries and are subject to import duties and taxes. The pharmaceutical industry in oncology is a challenge for SUS managers, due to constant technological innovation, making it necessary to constantly carry out economic evaluation studies (pharmacoeconomics) to acquire those that best serve the population (TORRES, 2010).

Another cost item we need to look at is inputs. A total of R$66,895.63 was spent in this sample in 2012, while R$52,514.42 was spent in 2006. Of these, a portion corresponds to the inputs spent on preparing and administering medicines, reminding us of how much this fraction of the procedure means in terms of labour for the nurse. Paralleling this with the hours spent on the related procedures, almost 1000 hours are spent on this part of the nursing service alone (SILVINO, 2008).

Inputs were the only item where the figures were higher in 2012 than in 2006. This can be explained by the cost of infection prevention procedures, since the complexity of the patients was lower. This difference in input costs, which totalled R$14,381.21, had a positive impact on the final result in terms of infectious complications. In terms of human resources, the nursing hours compared to 2012 were not significantly different. The impact of this cost on the transplant, however, was significant for 2006, when the total costs for the same year are evaluated. The many hours spent on care does not necessarily mean efficiency. In 2006, nursing work was carried out using the nursing process (parts of it) with no theoretical orientation and no methodology for carrying out care. Nursing work needs to be methodologically orientated in order to avoid rework. How were the nursing hours spent? According to Lefreve, the use of the nursing process has many benefits, such as reducing the incidence and length of hospital admissions by speeding up the diagnosis and treatment of health problems, creating a cost-effective plan, improving communication with the team, preventing errors and unnecessary repetitions and designing care for the individual and not just the disease.

We can say that the Systematisation of Nursing Care brought an important gain to the Bone Marrow Transplant Service, when we evaluate the year 2012 in relation to 2006, because in addition to showing reduced costs, it proves what the literature says about its implementation, which claims to obtain better results for the health of those who underwent the procedure in 2012 (AMANTE et al, 2009).

5.4 COST-MINIMISATION OF ALLOGENEIC BONE MARROW TRANSPLANTATION: THE INFLUENCE OF SAE ON TRANSPLANT COSTS

In order to carry out the cost-minimisation analysis of allogeneic stem cell transplantation with and without the implementation of SAE, we used the decision tree model and to compose it, data on the costs of transplantation in 2006 and 2012 as shown in the tables listed in item 5. 2, data on the complications associated with HSCT (table 19), obtained from the available literature and data on the cost of treating HSCT complications (table 20).Based on the data provided, a cost-minimisation analysis of related allogeneic HSCT was carried out in order to find out the least costly option in the two years listed, where in one year SAE was not in place and in the other it had already been implemented, as shown in figure 3:

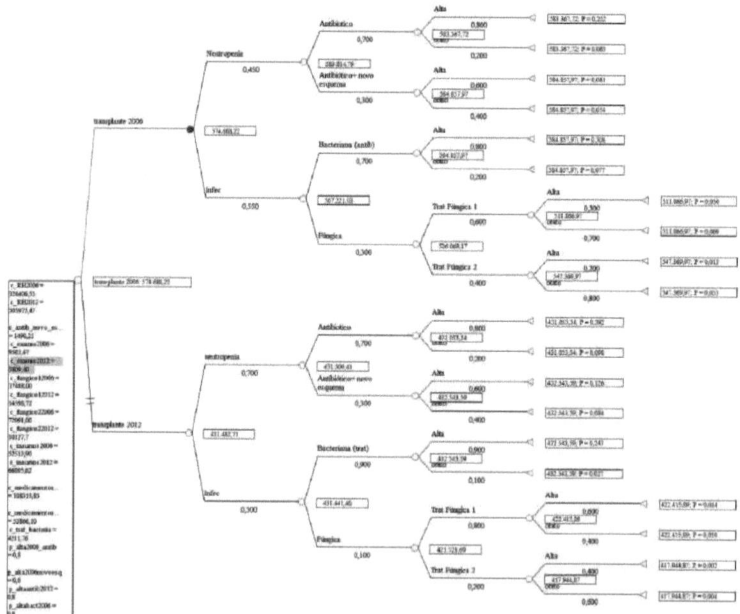

Figure 3 - Decision Tree

50

The cost-minimisation analysis of the transplant with the Systematisation of Nursing Care in 2012 showed a lower total cost when compared to the cost of the transplant carried out without the Systematisation. To assess the robustness of the cost-minimisation analysis, a series of one-way sensitivity analyses were carried out. To this end, all the cost variables included in the cost-minimisation calculation were modified by +/- 20% of their estimated base value.

The cost item that had the biggest impact in 2006 was medication, with a total of R$198,353.85. This is due to the high cost of medicines and also the incidence of infections with treatments for febrile neutropenia (R$17437.45), bacterial infection (R$4211.76) and fungal infection (R$110,479.82). Patients undergoing HSCT are at high risk of complications, requiring the nursing team to provide qualified, accurate and targeted care. Adequate healthcare provision is directly related to the quality and quantity of human resources available (MARTINS, 2013; ROTHBERG et al. 2005).

Drawing up the Systematisation of Nursing Care is one of the ways nurses can apply their technical-scientific and human knowledge to patient care and characterise their professional practice, helping to define their role (SPERANDIO and EVORA, 2008).

The time spent on labour in 2006 was much higher than in 2012, but despite the studies in this area of sizing, we know that quantity isn't everything. Few studies in the nursing field address the influence of clinical nursing protocols on the total costs of any treatment, whether in public or private institutions. Some of the aforementioned authors who have carried out studies on this subject claim that the use of the Systematised Nursing Care methodology brings better results in terms of quality of care and patient-related outcomes (TRUPPEL, 2009; LIMA and CASTILHO, 2012; MATA and SCHUTZ, 2012; TOLENTINO and SCHUTZ, 2013).

Knowing the impact of bone marrow transplant costs for the SUS and the need for more beds available for treatment, its effectiveness must be guaranteed by planning the work of nurses, as well as good cost management. Among the comorbidities studied in transplantation, it is known that the cost of treating infections adds to the total cost of the procedure, as well as requiring more nursing hours due to the complexity of the infected patient (COUTINHO, 2009).

In 2012, the cost of nursing labour was R$305,483.60, lower than in 2006 when it was R$324,406.55. This figure for 2012 was also significant and more cost-effective. The significance of this fact lies in the fact that nursing hours are spent targeting the most frequent risk diagnoses, thus working to prevent complications, which optimises clinical results and reduces total hospital costs (MAGALHÃES, 2005). Measures to prevent infections and care for patients of this nature include, among others: following febrile neutropenia protocols, catheter care, hand washing and maintaining a protective isolation environment. It is also worth highlighting the importance of educating patients and their families, as well as the healthcare team, in order to generate behaviours aimed at minimising the potential risks of infection (THUNGJAROENKUL et al, 2007).

We saw a much lower cost for medicines, R$53,866.10, as well as for tests, R$5,809.40. With regard to supplies, this was the only item that showed a higher cost compared to 2006 and is due to the investment in prevention techniques and the maintenance of protective isolation, as well as contact precautions, which justifies this higher expense. However, if we compare the cost of inputs with the cost of drugs to treat infections, the latter far exceeds the former (COUTINHO, 2009).To better visualise the data, we present in the figure below the incremental cost, which according to Nita (2010) is defined as the difference in costs when producing the same result from two different types of health intervention, both of which result in the same outcome.

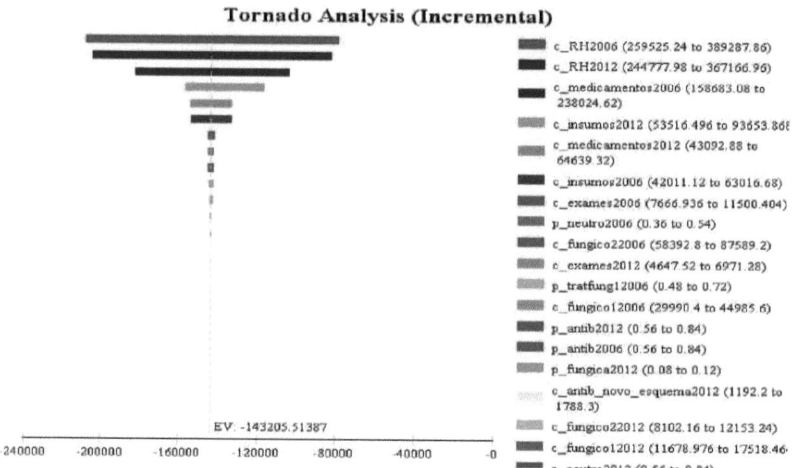

Figure 4 - Tornado diagram representing the incremental cost related to HSCT during the years 2006 and 2012

The tornado diagram shows the weight of the variables in the total cost of the transplants carried out in 2006 and 2012. The difference in costs between the two interventions results in the incremental cost (R$143,205.51). This amount could fund, for example, another allogeneic HSCT or two autologous transplants. We can also see that the weight of human resources in 2006 had the greatest impact on total costs, although it was not cost-effective as the results of the procedure in 2006 had greater morbidity and overall mortality rates were higher. The year 2012 had an important weight in terms of human resources, which, in relation to the results of the transplant in 2012, saw not only less morbidity, but also better survival rates. In 2006, the complexity of the patients was greater, requiring more nursing hours.

Drug costs in 2006 had a greater impact than in 2012, with prolonged and expensive treatments for fungal infections. The results did not show better effectiveness. In third place in the impact on total costs we see inputs spent in 2012. Expenditure on inputs can be interpreted in two ways: due to the high complexity of the patient, which required higher expenditure (in the case of 2006), or consumption justified by infection prevention measures in patient handling, as was the case in 2012, with lower infection rates. We can say that the total cost of the first intervention and the total cost of the second intervention are not proportional to their respective outcomes, as their magnitude was different.

We can see that in the sample studied, the costs and probabilities of intervention were compared. If we consider the possibility of alternative use of the necessary productive resources (opportunity cost), we could think of more outpatient care, outpatient transplantation, training and reform aimed at improving the service provided.

CHAPTER 6

FINAL CONSIDERATIONS

This study is a pioneer in economic evaluation relating SAE and stem cell transplantation. It sought to contribute to knowledge about the cost of haematopoietic stem cell transplants at a Federal Public Oncology Hospital located in Rio de Janeiro. With the implementation of the Systematisation of Nursing Care, it became possible to effectively apply the nursing process in all its stages, and we realised its importance not only from the point of view of health care for patients undergoing transplantation, but also from an economic point of view.

There was a scarcity of studies on economic evaluation by nurses in Brazil. The United States was the country in which the most studies on economic analysis by nurses were found. The experience of this work opens up new horizons for nursing research in HSCT, of whatever type, where various nursing procedures and protocols can be evaluated in a comparative way, calculating their costs and their effectiveness in service. It is considered necessary, both from the point of view of public management and for the benefit of the nursing profession, to continue studies that seek to validate the effectiveness of clinical protocols with a view to economic analysis. This study brings to light many needs within the bone marrow transplant service, such as comparative studies in the area of dressings, the cost of long-term follow-up of patients with graft-versus-host disease, and other issues within the area of transplantation that are relevant to the SUS, with the intention of guiding better technology, budget and reimbursement options. The analysis of the cost-minimisation of allogeneic HSCT showed that the cost with the implementation of the SNC is R\$ 432,543.59 lower and more effective, which is important data that legitimises our practice, while without the use of the SNC it is R\$ 584,858.49 higher, presenting important data for reflection on nursing work methodologies.

Many difficulties were encountered in the construction of this study, such as the delay in the evaluation process with the Research Ethics Committee via the Brazil platform, the lack of up-to-date published data on institutions that carry out transplants in Brazil, and especially the difficulty in obtaining cost data from the institution that served as the field for the study. It is important to emphasise that other factors may have influenced the results, such as changes in conditioning protocols and the greater availability of drugs with lower market prices. This study has contributed to the literature on bone marrow transplantation in Brazil, providing cost data that no other centre had done before. From the researcher's perspective, this study brought an immeasurable gain in the appropriation of knowledge in the area of health economics, giving a macro view of hospital cost management. For day-to-day practice, she became a multiplier of the cost concept, in the sense of generating better use of the resources available in the care area, and brought several reflections on the logistics of supplying inputs. With regard to the Systematisation of Nursing Care, we can evaluate this care methodology from various angles, and we have found in economic evaluation a new field of study that can greatly contribute to valuing nurses and impacting on their peers and the multi-professional team with regard to its efficiency.In the light of this study, it is important to emphasise the commitment of the Department of Fundamental Nursing, which has worked for the professional development of nurses by developing the health technology assessment laboratory, where ongoing research projects are discussed and scientific work is carried out in the area of health economics, a field that is still little explored by the health professional who does most of the management work: nurses.

REFERENCES

ABURDENE, P.; NAISBITT, J. **Megatrends for women**. 2 ed. Rio de Janeiro: Rosa dos Tempos, 1994. Chap. 5, p. 159-198. In: **Sistema Único de Saúde- SUS**: dados epidemiológicos; assistência médica. Brasília: MS, 1993.

AGIHO. **Management of sepsis in neutropenic patients**: 2014 updated guidelines from the Infectious Diseases Working Party of the German Society of Hematology and Medical Oncology.

ALMEIDA, M. H. **Hospital costs in nursing**. Rio de Janeiro: Cultura Médica, 1984.

AMANTE, L. N.; ROSSETTO, A. P.; SCHNEIDER, D. G. **Sistematização da assistência de enfermagem em unidade de terapia**. São Paulo: Revista da Escola de Enfermagem da USP, v. 43, n. 1, p. 54-64. 2009.

ARAÚJO, N. F. A. **Avaliação das políticas de transplantes de órgãos e tecidos no estado do Ceará**. 2008. Ceará State University: Thesis (Professional Master's Degree in Planning and Public Policies), 2008.

BACIGALUPO, A.; BALLEN, K.; RIZZO, D. et al. **Defining the intensity of regimens**: working definitions. Biol blood marrow transplant, v. 15, n. 12, p. 1628-1633 dec, 2009.

BACKERS, DS; SCHWARTZ, E. **Implementation of the systematisation of nursing care: challenges and achievements from the managerial point of view**. Ciência, cuidado e saúde, v. 4, n. 2, 2005.

BALLARD, B.; MITCHEL, K. **Renal and Hepatics effects**. In: WHEDON MB, WUJCIK D. **BLOOD AND MARROW STEM CELL TRANSPLANTATION**: PRINCIPLES, PRACTICE AND NURSING INSIGHTS. 2 nd ed. Boston: Jones and Bartlett Publishers, 1997, p. 298-325.
BEARMAN, S. L. **The syndrome of hepatic veno-occlusive disease after marrow transplantation**. USA, Denver: Blood, v. 85, n. 11, p. 3005-20, 1995.

BOUZAS, L.F.S. **Bone marrow transplantation in paediatrics and umbilical cord transplantation.** Ribeirão Preto: Medicina, n.33, p. 241-263, Jul. 2000.

BRASIL. **Para entender a gestão do SUS.** National Council of Health Secretaries. Brasília: CONASS, p. 248, 2003.

BROKEL, J. H. C. **The value of nursing diagnoses in electronic health records.** In: NANDA NURSING DIAGNOSIS: DEFINITIONS AND CLASSIFICATION 2007-2008/NORTH AMERICAN NURSING DIAGNOSIS ASSOCIATION; translated by Regina Machado Garcez. Porto alegre: Artmed. p. 343, 2008.

CAMPOLINA, A.G. Quality of life and utility measures: clinical parameters for health decision-making. **Revista Panamericana de Salud Pública,** Washington, v. 19, n. 2, feb. 2006.

CARVALHO, E. C.; BAHION, M. M. **Processo de enfermagem e sistematização da assistência de enfermagem.** UFG: Electronic Journal of Nursing, 2009.

COSTA, B. G.; MOTTA, I. J.; FARIAS, A. R. O. **Gestão econômica contribuindo para identificar oportunidades de melhoria em serviço público de hemoterapia.** Rio de Janeiro: Revista Brasileira de Hematologia e Hemoterapia (impresso), v. 29, p. 401-401, 2007.

INTERNATIONAL COUNCIL OF NURSES. **Quality, Costs and Nursing.** Geneva: Paper presented at the International Nurses Day, 1993.

COUTINHO, A. P. **Economic evaluation of the implementation of a protective environment unit for patients undergoing bone marrow transplantation.**
UFRGS: Master's dissertation, 2009.
CUNHA SMB; BARROS ALBL. **Analysis of the implementation of the systematisation of nursing care according to Horta's conceptual model.** Revista brasileira de enfermagem: v. 58, n. 5, p. 68-72, 2005.

DÓRO, M. P; PASQUINI, R. **Bone marrow transplant: a biopsychosocial confluence.** Curitiba: Iteração, v. 4, p. 39 - 60, jan - dec, 2000.

FERREIRA, PL. **Descriptive and inferential statistics.** Brief Notes, Portugal: 2005.

FRANCISCO, I.; CASTILHO, V. **Nursing and cost management.** Revista da escola de enfermagem da USP, v. 36, n. 3, p. 240-244, 2002.

FREIFELD, A. G; BOW, E. J.; SEPKOWITZ, K. A. **Clinical practice guideline for the use of antimicrobial agents in neutropenic patients with cancer.** Clinical Infections, 2011.

GEORGE, J. B. et al. **Nursing theories**: the foundations for professional practice. 4 ed. Porto Alegre: Artmed, 2000.

HAMERSCHLAK, N.; BOUZAS, L. F.; SEBER, A.; SILLA, L.; RUIZ, M. A. **Guidelines of the Brazilian society of bone marrow transplantation**. SBTMO, 2012.

HELAL; BYZIN, A.; REROLE, J. P.; MORETON, E. ;KREIS; BRUNEEL-MANZER, M. F. **Acute renal failure following haematopoietic cell transplantation**: Incidence, outcome and risk factors. Paris, France:Saudi centre for organ transplantation, 2011.

HAEMATOLOGY: FUNDAMENTALS AND PRACTICE. 1 ed. São Paulo: Ateneu, p 419 - 429, 2002.

INCA. **Nursing actions for cancer control, a proposal for teaching-service integration**. 3 ed. 2008. INCA, 2014. Incidence of Cancer in Brazil. **Ministry of Health; José Alencar Gomes da Silva National Cancer Institute**. 2014 estimate.

KARGAR, M. et al. **The prevalence of antifungal agents administration in patients undergoing allogeneic haematopoietic stem cell transplantation**: a retrospective study. Theran, Iran: International journal of haematology oncology and stem cell research, 2013.

KASSIER J. P.; MOSKOWITZ A. J.; LAU J.; PAUKER S. G. **Decision analysis: a progress report**. Ann: INTERNATIONAL MED, 33, 1987, Shlipak MG; v,106, n. 2, p. 275 - 291. Decision analysis. **In: Friedman, editor. Evidence-based medicine: a framework for clinical practice**. Rio de Janeiro: Editora Guanabara Koogan, p. 32-51, 2001.

KING, C.; HOFFAR, N.; MURRAY, M. **Acute renal failure in bone marrow transplantation**. USA: Oncol Nurs Forum v. 19, n. 9, p. 1327-35, 1992.

LIMA, A. F. C.; CASTILHO, V.; FUGULIN, F. **Cost of nursing activities performed more frequently in high dependency patients**. São Paulo: Revista Latino Americana de enfermagem, v. 20, n. 5, Sep 2012.

MACHADO, L. N.; CAMANDONI, V. O.; LEAL, K. P. H.; MOSCATELLO, E. L. M. **Transplante de medula óssea, abordagem multidisciplinar**. São Paulo: Editora Lemar, 2009.

MAGALHÃES, B. C. M.; PACHECO, C. R. M. **Nursing diagnoses of patients undergoing allogeneic bone marrow transplantation**: A case study. Porto Alegre: Revista gaúcha de enfermagem, v. 26, n. 1, p.

67-75, Apr. 2005.

MARGARIDO, E. S.; CASTILHO, V. **Aferição do tempo e do custo médio do trabalho da enfermagem na consulta de enfermagem**. Revista da escola de enfermagem da USP, v. 40, n. 3, p. 427-433, 2006.
MARTINS, Q. C. S. **Nursing Activies Score (NAS) as an instrument for measuring workload in a haematology and haematopoietic stem cell transplant unit**. Porto Alegre: Doctoral Thesis, Oct. 2013.

MATA, V. E.; SCHUTZ, V. **Cost-minimisation analysis of hydrogel and papain dressings for clients with venous ulcers**. UNIRIO: Master's dissertation, 2012.

MEDICI, A. C.; MARQUES R. M. **Cost system as an instrument of efficiency and quality of health services**. In cadernos. São Paulo:FUNDAP, v. 16, n 19, p. 47-59, 1996.

MENDES, E. D. T. **Clinical and epidemiological profile of patients undergoing haematopoietic stem cell transplantation in the haematology department of the FMUSP clinical hospital from 2001 to 2008**. SP: Monograph for the end of the medical residency course at FMUSP, 2010.

MENDES, E. D. **Clinical and epidemiological profile of patients undergoing haematopoietic cell transplantation in the haematology department of the FMUSP Hospital de Clínicas from 2001 to 2009**. SP: TCC Medical Residency, 2010.

NAKAO, J. R. **Enfermagem no processo de gestão econômica dos serviços de saúde: limites e possibilidades**. Brasília: Revista Brasileira Enfermagem, v. 52, n. 2, p. 223-232,1999.

NICOLAU, J. E. **Short-stay allogeneic bone marrow transplantation**. USP: Thesis for the title of Doctor of the Faculty of Medicine of the University of São Paulo, 2004.

NITA, M. E.; SECOLI, S. R.; NOBRE, MRC et al. **Health technology assessment**. São Paulo: Artmed, 2010.

NEVES, M. L. V. S. **Technique for the preparation and administration of intravenous medication by nurses with a view to haematopoietic stem cell transplantation**. Capstone presented to INCA to obtain a specialist degree, 2008.

NOBRE, M. R. C.; BERNARDO, W. M.; JATENE, F. B. **Evidence-based clinical practice. Part I: well-constructed clinical questions**. Revista Assoc Med Brasil, v. 49, n. 4, p. 445-9, 2003.

NUCCI, M. **Infections in patients with haematological neoplasms**: diagnosis, treatment and prevention.

In: ZAGO M. A.; FALCÃO, R. P.; PASQUINI, R. **Haematology**: fundamentals and practice. 1 ed. São Paulo: Ateneu, p. 419-29; 2002.

NUCCI, M. **Infections in bone marrow transplantation Medicine**. TRANSPLANTE DE MEDULA OSSEA, 33 ed. Ribeirão Preto, p. 278-293, Jul-Sep, 2000.

OLAF P.; CAROLIN, B.; DIETER, B.; MAXIMILIAN C. et al. **Helmut Ostermann**. Published online: 29 April 2014.

ORTEGA, E. T. T. **Compendium of nursing in haematopoietic stem cell transplantation**. EDITORA MAY, 2004.

PADILHA, M. I. **A qualidade da assistência de enfermagem e os custos hospitalares**. São Paulo:Revista hospitalar de administração em saúde, v. 14, n. 3, p. 128-133, 1990.

PANTON, E. J. A.; COUTINHO, M. A.; VOLTARELLI, J. C. **Diagnosis and treatment of acute complications of haematopoietic progenitor cell transplantation**. Ribeirão Preto: Medicina, v. 33, p. 264-277 jul. 2000.

PASSWEG, JR.; BALDOMERO, H.; BREGNI, M. et al. **Hematopoietic SCT in Europe: data and trends in 2011**. Bone Marrow transplantation. Milan: p. 1-7, 2013.

PERES, H. H. C; ORTIZ, D. C. F. **Electronic health information systems and the nursing process**. In: GAIDZINSKI, R. R.; SOARES, A. V. N.; LIMA, A. F. C.; et al. **Diagnóstico de enfermagem na prática clínica**. Porto Alegre: Artmed, p.338-53, 2008.
PENACK, O.; BECKER, C. et al. **Management of sepsis in neutropenic patients**: 2014 updated guidelines from the infections diseases working party of the german society of haematology and medical. Ann: Hematol, v. 93, p. 1083-1095, 2014.

POLÍTICA NACIONAL DE GESTÃO ESTRATÉGICA E PARTICIPATIVA, Brasília, DF, p. 221-22, 2007.

REDOME. **National Cancer Institute & Cancer Foundation**. 2009.

REPETO, M. A.; SOUZA MF. **Evaluation of the systematisation of nursing care in a university hospital**. PhD thesis UNIFESP, 2003.

BRAZILIAN TRANSPLANT REGISTRY; SBTMO, 2012.

RIZZO, J. D.; VOGELSANG, G. B.; KRUMM, S.; FRINK, B.; MOCK; V; BASS, E. B. **Outpatient-based bone marrow transplantation for haematologic malignancies: cost saving or cost shifting?** Baltimore: Journal clinical oncology: Johns Hopkins University, v. 17, n. 9, p. 2811-18, 1999.

ROTHBERG, M. B.; ABRAHAM, I.; LINDENAUERP K.; ROSE, D. N. **Improving Nurse-to-Patient Staffing ratios as a cost-effective safety intervention**. Medical care, v. 43, n. 8, aug. 2005.

SANTOS, D. S.; CARVALHO, E. C. **Cost analysis in nursing: integrative review**. Rio de Janeiro: Online Brazilian Journal of Nursing, v. 7. n. 3, 2008.

SCHUTZ, V. **O cuidado da enfermeira no mercado da saúde**: um estudo sobre o custo e o preço do processo de cuidar. Federal University of Rio de Janeiro, 2007.

SECOLI, SR; PADILHA, KG; LITVOC J, MAEDA, ST. **Pharmacoeconomics: an emerging perspective in the decision-making process**. Ciência e saúde coletiva; v. 10 (suppl):287-296, 2005.
SEAUNEZ, H. **Project**: Institute of science and technology for cancer control. INCA, 2008.

BONE MARROW TRANSPLANT SERVICE. **Hospital de Clínicas**: Manual of nursing procedures. Curitiba: Federal University of Paraná, 2003.

SILVA, L. K. **Technological evaluation and cost-effectiveness analysis in health**: the incorporation of technologies and the production of clinical guidelines for the SUS. Ciência e Saúde coletiva, v. 8, n. 2, p. 501-520, 2003.

SILVA, L. K. health technology assessment: bone densitometry therapeutic alternatives in post-menopausal osteoporosis. **Cadernos de Saúde Pública**, RJ: v. 19, n. 4 p.987-1003. 2003.

SILVEIRA, R. C.; GALVÃO, C. M. **Nursing care and the**
Hickman: the search for evidence. Acta Paul, v.18, n. 3, p. 276 - 284, 2005.

SOAREZ, P. C.; SOARES, M. O.; NOVAES, H. M. D. **Decision models for economic evaluations in health**. Rio de Janeiro: Ciência e Saúde coletiva, v. 19, n. 10, p. 4209-4222, 2014.

SPERANDIO, DJ; EVORA, YDM. **Proposal for implementing the systematisation of nursing care in a semi-intensive care unit**. n; 8. Simpósio brasileiro de comunicação em enfermagem, May, 2002.

TALLO, F. S.; VENDRAME, L. S.; LOPES, R. D,; LOPES, A, C. **Tumour lysis syndrome: a review for the clinician**. Revista brasileira de clinica medica. v. 11, n. 2, p. 150-4, 2013.

TELLES, S. C. R.; CASTILHO. **Personnel costs in intensive care unit nursing care**. USP: Revista Latino Americana de Enfermagem, 2007, v. 15, n. 5, sep. - Oct.

TOLENTINO, A. C. SCHUTZ, V. **Analysis of costs associated with indwelling bladder catheterisation**. Recife: Integrative review, revol, v. 7(esp), p. 4251-60, May, 2013.

THUNGJAROENKUL, P.; CUMMINGS, G. G.; EMBLETON, A. **The Impact of Nurse Staffing on Hospital Costs and Patient Length of Stay: a Systematic review**. Nursing economic $: v. 25, sep-oct- nov, 2007.

TRUPPEL, T. C.; MEIER, M. J.; CALIXTO, R. C.; PERUZZO, S. A.; CROZETA, K. **Systematisation of nursing care in an intensive care unit**. Revista Brasileira de Enfermagem, v. 62, n. 2, p. 221-227, Mar-April 2009.

VERSA, G. L. G. S.; INOUE, K. C.; NICOLA, A. L.; MATSUDA, L. M. **Influence of nursing team sizing on the quality of care for critically ill patients**. Texto & Contexto Enfermagem, v. 20, n. 4, oct-dic, 2011.

VOLTARELLI, J.; STRACIEN, A. B. **Immunological aspects of haematopoietic stem cell transplantation**. Ribeirão Preto: Medicina, v. 33, p. 443-462, Oct. - Dec, 2000.

WOJTASZEK, C. **Management of chemotherapy-induced stomatis**. Clin J Nurs, Australia: v. 4, n. 6, p. 263-70, 2000.

ZAFRANI, L.; AZOULAY, E. **How to treat severe infectious in critically ill neutropenic patients?** BMC infectious diseases, v. 14, p. 512, 2014.

ZUNTA, R. S. B.; CASTILHO, V. **Billing for nursing procedures in an Intensive Care Unit**. São Paulo: V.19, 2011.

BRAZIL PLATFORM

2.1 **DETAILING THE RESEARCH PROJECT**

RESEARCH PROJECT DATA

Research Title: FINANCIAL IMPACT OF THE SYSTEMATISATION OF NURSING CARE (SAE) IN THE ALLOGENIC TRANSPLANTATION OF HEMATOPOETIC STEM CELLS

Researcher: Solange dos Santos Moragas Barbosa

Thematic area:

Version:4

CAAE: 20869513.6.0000.5285

Submitted: 17/04/2014

Proposing Institution: Federal University of the State of Rio de Janeiro - UNIRIO

Status: Approved

Current Project Location: Researcher in Charge

Main Sponsor:Self-financing

Posted Project Documents

Document Type	Situation	Archive	Post
Opinion Consubstantiated by POSTCODE	A	PB CONSUBSTANTIATED OPINION CEP 625267 E1.pdf	24/04/2014 22:29:55
Basic Project Information	A	PB BASIC INFORMATION 28100 5 E1.pdf	17/04/2014 16:17:52
REBEC interface	A	PB XML INTERFACE REBEC E1.xml	17/04/2014 16:17:52
Others	A	Doc0001.pdf	17/04/2014 15:59:24
Detailed Design	A	project update with summary 1 .docx	17/04/2014 15:35:31
Cover Sheet	A	unirio cover sheet.pdf	20/08/2013 16:14:36
Opinion Consubstantiated by POSTCODE	A	PB CONSUBSTANTIATED OPINION CEP 670268.pdf	CEP Consubstantiated Opinion

JUSTIFICATION FOR NOT USING THE TCLE

Request to the Research Ethics Committee to carry out the research

As a master's student at the Federal University of the State of Rio de Janeiro (UNIRIO), under the guidance of Professor Vivian Schultz, we are developing a research project that aims to identify the direct cost items of allogeneic haematopoietic stem cell transplantation (HSCT), value the direct cost items associated with haematopoietic stem cell transplantation and carry out a cost-minimisation analysis with and without Systematisation of Nursing Care (SNC) in patients undergoing HSCT, the title of which is Financial Impact of SNC in allogeneic haematopoietic stem cell transplantation. This research, if approved by this CEP, will use data obtained from the medical records of patients who underwent allogeneic HSCT in 2006 and 2012 at a federal oncology hospital in the city of Rio de Janeiro, in the bone marrow transplant centre, from September to November 2013.

We guarantee the confidentiality of the data collected and that identities will not be revealed at any time, in accordance with CNS resolution 196/96.

If you would like further information, please contact us at 99757725 Solange dos Santos Moragas Barbosa and 78955081 (Prof Dr Vivian) or at Avenida Pasteur, 296, Urca (UNIRIO Nursing School building).

In these terms, we ask for your favour,

Rio de Janeiro, 13 August 2013.

PRICE RELATIONS

Year 2006

A - Inputs	Unit	Cost in R$
40x12 needles	Unit	0,69
30x8 needles	Unit	0,07
Insulin needles	Unit	0,03
Almotolia	Unit	0,7
2-way tunneled catheter	Unit	400
Camera equipment	Unit	8,48
Equipment with flow control	Unit	18,5
Equipment without sigh	Unit	0,48
Equipo w/ sigh	Unit	4,49
Pump equipment	Unit	64,34
Blood kit	Unit	2,6
Transfer equipment	Unit	7,76
Adhesive plaster	Unit	3,09
Blood culture bottle	Unit	20
Gauze pack	Package	0,38
Lancets	Unit	0,29
Surgical gloves	Par	0,31
Procedure gloves	Box	14
Micropore	Unit	3,6
Nebuliser	Set	96,35
Plug	Unit	8
Polifix 4-way	Unit	3,74
Tap	Unit	0,61
Collection tube with gel	Unit	0,4
Edta collection tube	Unit	0,2
Scalp 21	Unit	0,42
Scalp 23	Unit	0,45
Syringe 01 ml	Unit	0,15
03 ml syringe	Unit	0,11
05 ml syringe	Unit	0,09
10 ml syringe	Unit	0,32
20 ml syringe	Unit	0,52
60 ml syringe	Unit	3,9
B - Medicines	**Unit**	**Cost in R$**

Acyclovir cp	Cp	0,1
Acyclovirfa	Fa	3,72
Folinic acid cp	Cp	0,41
Folinic acid fa	Fa	6,18
Adrenaline	Am	0,29
Age bottle 100 ml	Fr	4,9
Water for injection 20 ml	Am	0,14
Albendazole 200 mg	Cp	0,08
Albumin	Fr	43,58
70% alcohol 1000 ml	Fr	4,17
Amikacin 500 mg	Am	0,76
Amiodarone	Am	0,68
Amiodarone	Cp	0,13
Amphotericin b 50 mg	Fa	10,2
Amphotericin lipid comp	Fa	1105
Amphotericin cream	Tube	3,35
Liposomal amphotericin	Fa	360,93
Anlodipine 5 mg	Cp	0,02
Atropine	Am	0,21
Bacitracin	Tube	0,91
Bactrim amp	Am	0,47
Bezafibrate 400 mg	Cp	1,76
Baking soda 250 ml	Fr	5,54
Sodium bicarbonate amp	Am	0,34
Bromopride amp	Am	0,87
Captopril 12.5 mg	Cp	0,02
Captopril 25 mg	Cp	0,02
Carmellose eye drops	Fr	17,8
Caspofungin 50 mg	Fa	1.400,00
Cefepime 2g	Fa	15,95
Ceftriaxone 1g	Fa	2,11
Cyclosporine amp	Am	60,15
Ciclosporin caps	Cp	4,05
Ciclosporin oral sol	Fr	227,06
Cipro cp	Cp	0,16
Cipro bottle	Fr	3,12
Clarithromycin 500 mg	Fa	12
Potassium chloride 10% 10 ml	Fa	0,13
Sodium chloride 0.9% 20 ml	Fa	0,16
Sodium chloride 20% 10 ml	Fa	0,13
Potassium chloride syrup	Fr	1,58
Chlorhexidine 0.5% alc 11	Fr	4,9
Chlorhexidine degerm 11	Fr	13,6
Complex b	Am	0,28
Haemaceous cone	Bo	75,1
Platelet cone	Bo	75,1
Dexamethasone cream	Tube	0,7
Dexamethasone fa	Fa	0,53
Diazepan amp	Am	0,35
Diazepan 5mg	Cp	0,56
Diazepan 10mg	Cp	0,02
Dipyrone amp	Am	0,25
Dobutamine	Am	0,56
Dolantina	Am	0,47
Dopamine	Am	0,55

Phenergan	Am	0,55
Phenytoin 100 mg	Cp	0,3
Phenytoin amp	Am	0,8
Fenoterol drops	Fr	1,56
Fentanyl 10ml	Fa	2,76
Fentanyl25 mcg patch	Unit	26,48
Fluconazole cp	Cp	0,27
Fluconazole bottle	Fr	3,26
Phosphate pot 10% 10 ml	Am	0,37
Furosemide amp	Am	0,23
Furosemide cp	Cp	0,02
Ganciclovir 500 mg	Fa	34,2
Gcsf 300mg	Fa	30,5
Glucose 25% amp	Am	0,15
Glucose 50% amp	Am	0,15
Calcium gluconate	Am	0,99
Haloperidol amp	Am	0,5
Heparin fa	Fa	2,01
Hydrocortisone 100 mg	Am	1,21
Hydrocortisone 500 mg	Fa	3
Hydrochlorothiazide cp 25 mg	Cp	0,02
Hyoscine	Am	0,35
Imipenem 500 mg	Fa	21,65
Immunoglobulin 5 g	Fr	150
Insulin nph	Fa	35,5
Regular insulin	Units	35,5
Ipratropium drops	Fr	1,28
Lactulose	Fr	11,68
Linezolid 600 mg	Bo	171
Mannitol 20 % 250 ml	Fr	1,84
Meronen 1 g	Am	89,9
Meronen 500 mg	Fa	73
Metrotexate 50 mg	Fa	7,48
Mycophenolate 500 mg	Cp	6,12
Midazolan 15 mg 3 ml	Am	1,2
Midazolan 50 mg	Am	1,97
MorfinalO mg	Am	0,74
Noradrenaline 4 ml	Am	3,26
NorestironalO mg	Cp	0,37
Npt	Bo	80
Mineral oil	Fr	1,08
Mineral oil bottle	Fa	1,11
Omeprazole 20 mg	Cp	0,08
Omeprazole 40 mg	Fa	2,43
Zinc oxide	Tube	1,63
Paracetamol	Cp	0,04
Plasma	Bo	75,1
Plasil	Am	0,2
Prednisone 20 mg	Cp	0,06
Will provide	Cp	1,8
Ringer's lactate 500 ml	Fr	0,79
S glyc 10% 500 ml	Fr	1
Setux	Fr	5,03
Simethicone	Fr	0,58
Solumedrol 500 mg	Fa	16,4

Physiological saline 0.9% 50 ml	Fr	1,41	
Physiological saline 0.9% 100 ml	Bo	1,43	
Physiological saline 0.9% 250 ml	Bo	1,51	
Physiological saline 0.9% 500 ml	Fr	1,78	
Physiological saline 0.9 % 1000 ml	Fr	1,99	
Glycine serum 5% 50 ml	Fr	1,48	
Gly Serum 5% 100ml	Bo	1,5	
Gly Serum 5% 250ml	Bo	1,57	
Glycine serum 5% 500 ml	Fr	1,72	
Glycine serum 5% 1000 ml	Fr	1,47	
Magnesium sulphate 10% 10ml	Fa	0,21	
Sulfadiazine cream	Tube	2	
Tacrolimus amp	Am	294,43	
Tramadol 100 mg	Am	0,54	
Tridil 50mg	Am	19	
Ursacol 150 mg	Cp	1,85	
Vancomycin 500 mg	Fa	3,67	
Vitamin k	Am	0,55	
Voriconazole 200 mg	Cp	12,5	
Voriconazole 200 mg	Fa	604,76	
Zofran 8mg	Am	0,79	

Year 2012

A - Inputs	Unit	Cost in R$
30x8 needles	Unit	0,04
40x12 needle	Unit	0,05
Insulin needles	Unit	0,03
Eva bag 1000	Unit	10,92
Film dressing	Unit	8,9
Common equipment with sigh	Unit	4,23
Pump equipment	Unit	91
Blood kit	Unit	2,6
Transfer equipment	Unit	9,4
Doseflow equipment	Unit	15,5
Microdrop tube	Unit	9,55
Adhesive plaster	Unit	4,27
Hgt tapes	Unit	0,22
Gauze pack	Package	1
Lancet	Unit	0,3
Surgical glove 7.5	Par	0,65
Procedure gloves m	Box	0,12
Micropore	Unit	2,96
Polifix 4-way	Unit	3,84
Plug	Unit	4,34
Syringe 01 ml	Unit	0,077
03 ml syringe	Unit	0,15
05 ml syringe	Unit	0,11
10 ml syringe	Unit	0,09
20 ml syringe	Unit	0,52
60 ml syringe	Unit	3,71
Blood culture tube	Unit	20
Collection tube with gel	Unit	0,31
Edta collection tube	Unit	0,21
B - Medicines	**Unit**	**Cost in R$**
Aas cp1OOmg	Cp	0,0195

Acyclovirfa	Fa	1,57
Age bottle 100 ml	Fr	10,85
Distilled water amp 20 ml	Fa	0,1
Albendazole 200 mg	Cp	0,1
Albendazole cp 400 mg	Cp	0,19
Albumin fr	Fr	59,6
Amiodarone	Am	0,75
70% alcohol 100 ml	Fr	0,7
Allopurinol cp 100 mg	Cp	0,02
Amikacin 500 mg		0,63
Amiodarone cp 200 mg	Cp	0,07
Liposomal amphotericin	Fa	718,43
Amphotericin ointment	Tube	5,36
Anlodipine cp 5 mg	Cp	0,02
Atenolol cp 50 mg	Cp	0,01
Atg 25 mg	Am	385,48
Atropine amp	Amp	0,17
Atrovent fr	Fr	1,18
Baclofen cp 10 mg	Cp	0,05
Bactrim amp	Amp	1,2
Berotec fr	Fr	0,65
Sodium bicarbonate 250ml bottle	Amp	10,31
Bone marrow grant		290,17
Bromopride amp	Amp	0,44
Bussulfan cmp 2 mg	Cp	0,59
Cancidas 50 mg		556,26
Captopril 12.5mg cp	Cp	0,02
Captopril25mg cp	Cp	0,01
Carmellose fr 15 mg		37,77
Caverdilo cp 3.125 mg	Cp	0,065
Cefepime fa	Fa	2,25
Ceftriaxone fa	Fa	0,97
Cyclophosphamide fr 1000 mg	Fr	24,83
Cyclosporine amp	Amp	62
Ciclosporin oral sol	Fr	205,91
Cipro bottle	Fr	2,29
Clexane 40 mg		7,37
Clexane 60mg		10,89
Potassium chloride 10% 10ml fa	Fa	0,1
Sodium chloride 20% fa 10ml	Fa	0,1
Chlorhexidine 0.5% 100 ml	Fr	1,29
Chlorhexidine degerm 100 ml	Fr	0,97
Chlorpramazine am 25 mg	Am	0,8
Codeine cp 30 mg	Cp	0,51
b amp complex	Amp	0,62
Dexamethasone amp	Amp	0,35
Dexamethasone cream	Tube	0,4
Diazepan amp	Amp	0,36
Diazepan cp 5mg	Cp	0,02
Diphenhydramine amp	Amp	9,02
Diphenhydramine cp	Cp	0
Dipyrone amp	Amp	0,18
Dipyrone fr 10ml	Fr	0,28
Dobutamine amp	Amp	1,68
Epinephrine amp	Amp	0,26

Spiranolactone cp 100 mg	Cp	0,24
Phenobarbital amp	Amp	0,69
Fentanyl ad	Sticker	10,28
Fluconazole cp	Cp	0,18
Fluconazole bottle	Fr	4,09
Fludarabine fa 50 mg	Fa	354,65
Potassium acid phosphate fa 10 ml	Fr	1,23
Furosemide amp	Amp	0,17
Furosemide cp	Cp	0,01
Gsf	Fa	10,19
Glycerophosphate fa 20 ml	Fa	48
Glucose 25% amp	Amp	0,16
Glucose 50% amp	Amp	0,16
Calcium gluconate	Fa	0,9
Haloperidol amp	Fa	0,34
Haemaceae	Scholarship	75,1
Heparin fa	Fa	5,74
Hydralazine dr25	Cp	0,15
Hydrochlorothiazide 25 mg	Cp	0,01
Hydrocortisone 100 mg	Fa	0,67
Hydroxide alum fr 200 ml	Fr	1,05
Hyoscine amp	Amp	0,4
Ibuprofen fr 20 ml	Fr	7,7
Immunoglobulin 6 gr	Fa	63,58
Insulin nph	Fr	9,93
Regular insulin	Amp	10,03
Itraconazole cp 100 mg	Cp	0,3
L.thyroxine sodium 100 mg	Cp	0,08
L.thyroxine sodium 75 mg	Cp	0,09
Lactulose	Fr	5,27
Leucovorin fa 50 mg	Fa	6,21
Lidocaine ampoule	Amp	0,36
Linezolid 600 mg	Scholarship	199,84
Losartan cp 50 mg	Cp	0,06
Meronen 1g	Fr	25,4
Mesna amp 400 mg	Amp	3,94
Methylprednisolone fr 500 mg	Fr	8,81
Metroclopramide amp	Amp	0,19
Midazolan amp 10 ml	Amp	3,06
Midazolan cp 15mg	Cp	0,36
Morphine 10mg amp	Amp	0,35
Mupiroxine ointment	Tube	5,29
Nebacetin ointment 20 g	Tube	0,35
Nifedipine cp 20mg	Cp	0,14
Norepinephrine 6 mg amp	Amp	0,85
Norethisterone cp 10 mg	Cp	0,42
Npt	Scholarship	19,95
Mineral oil ml	Fr	0,99
Omeprazole 40 mg fa	Fa	2,09
Omeprazole cp 20 mg	Cp	0,039
Zinc oxide tube	Tube	1,14
Penicillin potassium g fa	Fa	1,18
Platelets	Scholarship	75,10/17,04
Prednisone 20 mg	Cp	0,04
Prednisone cp5 mg	Cp	0,02

Promethazine cp25 mg	Cp	0,6
Ringer's lactate	Fr	1,29
Rivotril cp 0.5	Cp	0
Rivotril oral sol	Fr	12,89
Simethicone fr	Fr	0,38
Simvastatin cp10 mg/40	Cp	0,89
Sorophysiol. 0.9% 20 ml	Fr	0,14
Sorophysiol. 0.9% 50 ml	Fr	1,15
Saline 0.9% 100 ml	Fr	1
Sorophysiol. 0.9% 250 ml	Fr	0,97
Sorophysiol. 0.9% 500 ml	Fr	0,88
Saline 0.9% 1000 ml	Fr	1,44
Glucose serum 5% 50 ml	Fr	1,4
Glucose serum 5% 100 ml	Fr	1,1
Serum glyc.5% 250 ml	Fr	1,05
Glucose serum 5% 500 ml	Fr	1,1
Serum glucose 5% 1000 ml	Fr	1,9
Magnesium sulphate 10% 10 ml	Fa	0,2
Tacrolimus amp	Amp	354,96
Thyroxine cp 25 mg	Cp	0,01
Tramai amp 100 mg	Amp	0,41
Tylenol cp 500 mg	Fr	0,02
Ursacol cp 150 mg	Cp	1,23
Vancomycin 500 mg	Fa	2,2
Voriconazole 200 mg fr	Fr	646,55
Zofran 8mg amp	Amp	0,44
C - Exams	**Unit**	**Cost in R$**
Antigenemia for cmv	Unit	20
Calcium	Unit	0,35
Creatinine	Unit	0,12
Phosphatase	Unit	0,23
Phosphate	Unit	0,16
Blood count	Unit	7
Galactomannan	Unit	30
Glucose	Unit	0,12
Magnesium	Unit	0,26
Serum csa level	Unit	33
Tacrolimus serum level	Unit	65
Per	Unit	1,35
Potassium	Unit	0,29
Sodium	Unit	0,29
Tgo	Unit	0,19
Tgp	Unit	0,26
Ldh	Unit	0,41
Urea	Unit	0,1

More
Books!

info@omniscriptum.com
www.omniscriptum.com
OMNIScriptum

Printed by Books on Demand GmbH, Norderstedt / Germany